T0350205

WHAT MAKES YOU
CLEVER

the puzzle of intelligence

WHAT MAKES YOU
CLEVER
the puzzle of intelligence

Derek Partridge
University of Exeter, UK

 **World Scientific**

NEW JERSEY · LONDON · SINGAPORE · BEIJING · SHANGHAI · HONG KONG · TAIPEI · CHENNAI

Published by

World Scientific Publishing Co. Pte. Ltd.
5 Toh Tuck Link, Singapore 596224
USA office: 27 Warren Street, Suite 401-402, Hackensack, NJ 07601
UK office: 57 Shelton Street, Covent Garden, London WC2H 9HE

Library of Congress Cataloging-in-Publication Data
Partridge, Derek, 1945–   author.
 What makes you clever : the puzzle of intelligence / by Derek Partridge (University of
Exeter, UK).
   pages cm
  Includes index.
  ISBN 978-9814513036 (hardcover : alk. paper) -- ISBN 978-9814513043 (pbk. : alk. paper)
  1. Artificial intelligence. 2. Machine learning. I. Title.
  Q335.P3546 2014
  006.3--dc23
                                                                          2013051172

British Library Cataloguing-in-Publication Data
A catalogue record for this book is available from the British Library.

Cover art by Ng Chin Choon, with the assistance of Oliver Baker, oliverbacker.com and iCub face
photos courtesy of Lorenzo Natale of the Instituto Italiano di Tecnologia.

In-house Editor: Amanda Yun

Typeset by Stallion Press
Email: enquiries@stallionpress.com

Printed in Singapore

An eager, young Englishman found himself in prison and in isolation in a foreign land. Desperate for company he was elated to find the adjacent cell occupied by a man who spoke broken English. After some days of getting to know each other and generally exhausting the possibilities for chit-chat, the young man asked his fellow prisoner to teach him his native language. He was overjoyed to hear, "I will do that, but you must ask me what you want to know. I will not repeat myself. I will answer all your questions once, and once only."

"Okay," the young man agreed, "I'll write down everything on my cell wall, so I'll never need to ask you anything twice."

And so the language lessons began. First, just simple words and common phrases which the young man assiduously wrote on his cell wall as he repeated them to his teacher. As the months went by, and verbs were conjugated and idioms explained, the Englishman became proficient. In between lessons, his teacher told him about the wonders of this distant country — beautiful mountains, delightful forests, charming towns and villages. The young man was entranced by what he learned. Within a couple of years he began to compose poems to celebrate the beauties of the land that he was now so alive to. In the evenings, he read them out loud to the obvious delight of his neighbour.

Eventually, he was released from prison. Naturally, the first thing he wanted to do was to visit his erstwhile instructor's country in order to see its wonders, which he knew so intimately, and to exercise his fluency in the language.

He duly arrived and was somewhat surprised by the unremittingly flat desolation he saw where mountains and waterfalls ought to be. He walked up to the first native he encountered, greeted him with a well-practised phrase, and launched into a fluent question about the surprising landscape.

The native stared in puzzlement. He had absolutely no idea what the young man was babbling on about; not a word made any sense. The Englishman had learned the detailed structure of a language that had no meaning because it was a total fiction: a complex of dynamic structures, devoid of meaning in the world at large.

About the Author

Derek Partridge is a Computer Scientist with a special interest in Artificial Intelligence and the design and development of IT systems. A PhD awarded by Imperial College, London in 1972 initiated nearly four decades of teaching and research at Universities around the world (England, the USA, Kenya, Malaysia, Chile, and Australia). With more than 100 scholarly publications to his name, Derek's books include *The Foundations of AI: a sourcebook* (edited with Y. Wilks), *The Seductive Computer: why IT systems always fail*, *A New Guide to AI* and *Computers and Creativity* (with J. Rowe). Since retirement, as an Emeritus Professor at the University of Exeter, UK, he divides his time between research and writing, and managing a private nature reserve on the edge of Dartmoor National Park in the county of Devon, England.

For Ethan and Milo,
a little light reading to put you to sleep

Preface

Science has been one of humanity's great triumphs: the Hadron Collider is revealing the fleeting infinitesimals of matter; the Hubble telescope probing deep space is reporting on the beginnings of time and its galactic consequences; Craig Venter is presenting the world with man-made life; the engine of life, the genome, has yielded up its structure and is beginning to reveal its mechanisms; and so it goes on, the relentless march of science laying bare all that it encounters.

There is, however, one glaring exception — intelligence. Despite decades of intense scrutiny, we have no coherent explanation of how it might work except in terms of the haziest descriptions. Scientists have very little idea what makes you clever. Curiously, however, in half a century of serious scientific study of intelligence we have been relentlessly assured by the scientists (enthusiastically backed by the media) that breakthrough is imminent. How long can science maintain that it is teetering on the brink of understanding intelligence? Why would this particular science, one branch of the objective enterprise par excellence, persist with overblown claims, and yet make such little real progress?

You and I, and all our friends, are intelligent; all non-human species fall far short of us in this regard. Sophisticated intelligence is the exclusive preserve of humanity (so we fondly believe): we can all communicate in a human language; we can reason about abstract ideas such as numbers; we manipulate and manage beliefs about the past, the present and the future as well as about other people and their beliefs.

Astrophysics, genomics and the world of sub-atomic particles are not subjects that most of us will ever get close to; we must accept uncritically what the scientists tell us. Again, intelligence is peculiar. Every one of us is an expert; so when scientists explain their findings, we ought to understand them and be able to evaluate many of their proposals. Working against the benefits of immediacy we have some difficulty in maintaining the objectivity of distance between our subject and ourselves, as Desmond Morris is at pains to point out in the introduction to his zoological study *The Naked Ape* (McGraw-Hill, 1967).

What about your brain, where does this organ come into the picture? It is the seat of intelligence; it is also disconcertingly similar to the brains of other animals that show no signs of real intelligence. That's another part of the puzzle.

Is the human brain an antiquated engine that puts a brake on intelligence? The basic brain mechanisms and structures evolved over millions of years; intelligence emerged in the last few tens of thousands. Is intelligence a Formula 1 concept struggling to run on a Model T chassis and engine with just fuel injection lately added on? Or has evolution bestowed upon us a finely tuned optimal compromise between conflicting demands? To remember everything would be disastrous, as would remembering nothing.

In either case, can we extract the essence of intelligence, and program it into a modern computer? Surely then science would have intelligence thoroughly pegged. It would have an understanding sufficient to manipulate the human version, and perhaps even to create a super-intelligence.

The Deep Blue computer beat world chess champion Garry Kasparov in 1997, and in 2011 the Watson computer beat the two best human players in the question-answering TV game show *Jeopardy*. There are machines that can learn, cameras that can "see" and robots that can chat as well as exhibit many other human capabilities. Has so-called

Artificial Intelligence arrived? No it hasn't. Despite the regular TV cameos of robots behaving persuasively, science is still just scratching the surface. In *What Makes You Clever*, we see why science has yet to get a good grip on how intelligence works.

The difficulties can be reduced to ten maxims: stretching from the technical difficulties of the subject of intelligence (Maxims 1 to 4), to some unhelpful psychological predispositions stemming from the oddities of formal logic (Maxims 5 to 8), and the practicalities of computer model building (Maxims 9 and 10).

The ten maxims are:

1. A major technical difficulty: **the mind exhibits no joints for science to carve at.**
2. A testing problem: **intelligence admits no simple tests.**
3. A vicious circularity: **understanding based on a single example.**
4. An undeniable truth: **structure does not provide meaning.**
5. A genetic imperative: **if it's like me, it's intelligent.**
6. A logical conundrum: **we cannot prove, but we can disprove.**
7. A sad fact: **failure drives science forward.**
8. A measure of scientific progress: **by their fruits you will know them.**
9. A technological difficulty: **no significant computer program is completely understood.**
10. An unbelievable truth: **no computer will ever count 1, 2, ... exponentially many.**

The interaction of these ten maxims provides a basis for explaining why science is making such slow progress. They show why the workings of your day-to-day facility with survival in the world is still very much a mystery despite relentless reports of great strides forward in brain science, in cognitive science and in computer modelling of theories of intelligence, known as Artificial Intelligence or AI.

The science of cognition is probing our very humanness. What do you know? How do you reason? What do you tend to forget? How do you manage to chat? In sum: what makes you clever? To all these questions science is seeking answers, and you the reader have an indisputable claim to the relevant expertise. Intelligence is the one field of scientific enquiry that we are all qualified to comment on. How does intelligence work — and not work? Your insights may be as good as those of anyone, and your questions just as pertinent once a basic appreciation of the science has been achieved. This book explains the science, and how it is struggling with the many puzzles of intelligence. Every reader is invited to evaluate the science.

The introductory sentence of the Forward to Richard Dawkins' blockbuster *The Selfish Gene* (Oxford University Press, 1976) is: "The chimpanzee and the human share about 99.5% of their evolutionary history, yet most human thinkers regard the chimp as a malformed, irrelevant oddity while seeing themselves as stepping-stones to the Almighty." Regardless of the significance we care to attach to it, this lacuna undoubtedly exists and science has yet to get a firm grip on what is happening at the human end. What accounts for humanity's "great leap forward"? On the basis of the ten maxims, *What Makes You Clever* explains the puzzles, exposes the hype and points to possible ways forward.

Derek Partridge
December 2013
Hennock, Devon, UK

Acknowledgements

The content of this book is a distillation of decades of researching, teaching and pondering the puzzle: I think, therefore what am I doing? Consequently, much of the credit for insights and assistance provided by others has been lost in the "eternal lines to time". Crucial neurons may have passed away, or maybe been hijacked by newer memories, who knows? This, itself, is a part of the puzzle.

Of more recent and direct involvement with the book that has emerged, some names are still reverberating insistently within my time-ravaged neural network: my recent ex-colleague, Antony Galton, and my not-so-recent ex-colleague, Yorick Wilks, have both read, commented on and discussed textual drafts as well as advised on the underlying ideas. From across the Atlantic Victor Johnston in New Mexico and Phil Holcomb have advised on the brain science aspects as did Ken Paap. Ken and I jointly developed the "red herring of recursion" basis for Chapter 12[1]. Marc Hauser kindly responded to my queries regarding his seminal "recursion" paper, with Fitch and Chomsky, in *Science*. Ed Keedwell at Exeter checked out my explanation of Genetic Algorithms, and we've agreed to differ a little here and there.

---

[1] Our joint idea was published as Partridge, D., & Paap, K. R. (2011, November). "Recursion Isn't Necessary, Much Less the Hallmark of Human Language" as a poster presented at the 52nd Annual Meeting of the Psychonomic Society, Seattle, WA, and a full paper by K.R. Paap & D. Partridge entitled "Recursion Isn't Necessary for Human Language Processing" is to be published in the journal *Minds & Machines*.

Angelo Cangelosi, an iCub robot researcher at the University of Plymouth, usefully rounded out my beliefs about the Europe-wide spectrum of iCub projects.

Thanks also go to Alasdair Forbes for whom the full scope of objective science is no more than a first step, perhaps just a stumbling block, on the long road to the really important questions. He did his best to direct my text towards increased semantic precision, and my mind towards the real problems of the human condition.

Peter Wiseman also refined some of my prose as well as exercised his scholarly expertise to verify my views about what the Romans did for the Gauls. Similarly, Peter Quartermaine added some inconspicuous, but nonetheless valuable, syntactic refinements. Oliver Baker, yet another non-scientist, pointed out where the prose of the scientist gave no quarter to the normal reader.

The opening story of pseudo-language learning is loosely remembered from a movie, but once more (and despite assiduous Googling) the relevant neurons stubbornly refuse to play ball. So, no further precision can be attached to this particular acknowledgement.

Finally, in this age of electronic publishing human competence all too often gives way to a supposedly adequate human-computer combination in which human expertise is deemed unnecessary. *World Scientific* have been a refreshing surprise. In particular, I would like to acknowledge the constant and ever-ready support from Amanda Yun, my editor, and the similar responsiveness and attention to detail of Juleen Shaw the copy-editor. Together, they have done sterling work trying to induce a scientist to write more plainly and simply (not to mention all the grammatical infelicities and oversights corrected).

Contents

෴ Chapter 1 ෴

A Singular Enigma

"We can't understand the plot of Downton Abbey *by taking apart the television."*[1]

I think, and you think. But *how* we manage to think is almost a total mystery to science. In sharp contrast to almost every other topic that scientists have tackled, your cleverness remains an enigma despite decades of scientific enquiry. From this book you will find out why your easy skills — such as chatting — have proved so difficult for science to understand. Your cleverness is not easily explained, but by following the scientific probing described in the following chapters you will learn a lot about this accomplishment that makes you unique in the vast tree of life. You will also learn about the nature of scientific enquiry itself.

Because we all think, every reader has a sound basis for assessing what the scientists say about "thinking". So what do they say?

Alan Turing, the British mathematician rightly famed for his contributions to the cracking of the German Enigma codes in the Second World War, was one scientist who grappled with the notion of "thinking".

Pushed to an early death at age 41 by the society he had done so much to save, Turing's fundamental contributions to computer science are less well known outside the world of computing. He defined the abstract model, the Turing Machine, upon which modern Information Technology (IT) still rests. He also constructed a

1

number of fundamental proofs pertaining to the scope and limitations of this technology.

In addition, Turing has a good claim to be the father of the quest to construct IT systems — essentially computer programs — that exhibit intelligence, the field of endeavour known as Artificial Intelligence (AI). Squarely in the middle of the 20th century he conjectured that computers, which hardly existed at the time, might be programmed to play chess and participate actively in other hitherto exclusively human activities.

Turing wisely side-stepped the idea of defining what constitutes thinking and instead proposed the Imitation Game, often known today as the Turing Test[2]. Turing's idea was that if a person is allowed to engage another unseen "system" (either another person or a suitably programmed computer) in unrestricted dialogue, and consistently mistakes the computer for another person, then it must be conceded that the computer is thinking. Passing the Turing Test is the proposed criterion of "thinking", in effect a definition of "intelligence", and is one that has stood the test of time, although it is not without its critics[3].

This is how the experiment goes: viewing yourself as the interrogator, your aim is to "spot" the computer system — is it system A or system B? You can ask whatever you like, but the computer system is, of course, under no obligation to be truthful.

> **A:** OK, I'm the person here. Ask me any question and I'll prove I'm the human.
>
> **Interrogator:** A, how many fingers on your left hand?
>
> **A:** Five, of course — well maybe four and a thumb.
>
> **Interrogator:** B, what do mangoes taste like?
>
> **B:** I've never tasted a mango so I cannot tell you.
>
> **Interrogator:** Both of you, what's 282 multiplied by 40?
>
> **A:** Sorry, I need a calculator for that one. I can't do sums.
>
> **B:** Hang on a minute… it's 11,240.

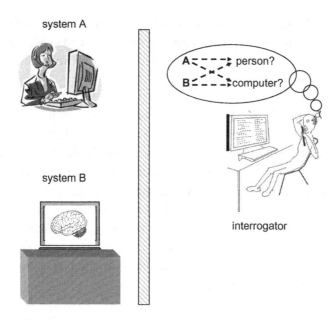

Within this experiment, you are as qualified as anyone else to play the part of either the interrogator or system A. How would you go about either of these roles? As system A you are aiming to convince the interrogator that you are human and quite distinct from the computer system beside you. Absolutely anything can be said, but what sort of things would you say? You would be trying to project "humanness"[4], but what are the signals of that?

You might focus on the various consequences of having a human body or interacting within a community of humans. In both situations, system B (which doesn't presumably have a human-like body nor live within a human community as a human) might be expected to reveal its artificiality.

Both roles bear much thinking about (as many commentaries on the Turing Test have explored) but produce precious few clear directives. Is system A the person in the above dialogue, or is it system B? A communicates like a human, but B got its multiplication wrong. It's a tough decision, indeed an impossible one with only a small snippet to go on, because **intelligence admits no simple tests** (Maxim 2) — a first collision with scientific practice. Hence, the Turing Test requirement of extended dialogue[5].

Could jokes be a key? What if the interrogator tells jokes, and then assesses the responses of system A and system B? Will the computer reveal itself by consistently failing to be amused? We don't know, but it's a plausible idea because the humour of joke appreciation appears to tap the deep subtleties of word meanings. Ranged against this strategy, an attempted humour probe can be deflected by a response, such as, "Just be serious, I am not in the mood for jokes." But any such quashing of potential for humour does require some recognition that humour is being attempted, unless launched as a pre-emptive strike. And so the possibilities go on...

It's an interesting puzzle that any of us can join in. We all have the capacity to engage fully; this central experiment is not the exclusive preserve of experts. Your insights are likely to be as good as anyone else's.

The underlying assumption of the Turing Test is: if you can hold up your end of a prolonged chat in English, then you must be thinking. This means that you are intelligent. Many scientists hold this to be so. Is it reasonable?

You can talk more or less effortlessly with many people in many different situations. Does this mean that you are always exercising your intelligence whilst chatting away? Apart from your knowledge of the language itself — what can be said and how it can be said — you use your knowledge of the particular situations under discussion, your knowledge of the person you're talking to, common sense knowledge, and knowledge of the world. That's a lot of knowledge that must be stored and organized, and integrated into your sentences. So, maybe extended and sensible conversation does imply intelligence? Turing would have it so. Would you?

4

Turing gave us a definition that we can work with but he was, of course, by no means the first luminary to single out human language as a key indicator of intelligent thought. Early in the 17th century, for example, the philosopher Descartes, famous for "I think, therefore I am", proposed the human capacity for language as a "means [to]... tell the difference between men and beasts":

No men are so dull-witted and stupid, not even imbeciles, who are incapable of arranging together different words, and of composing discourse by which to make their thoughts understood; and that on the contrary, there is no other animal, however perfect and whatever excellent dispositions it has at birth, which can do the same.[6]

Does an ability to chat seem reasonable as a definitive test of basic intelligence, in a word, cleverness? Your hunches are as valid as mine, and similarly not to be relied upon, but that does not mean they are worthless. Such insights do present us all with a basis for examining these elements of a scientific understanding of intelligence, provided we do not take them to be indisputable guides to truth.

Science suggests that extended and diverse dialogue is complex because it must involve management of similarly diverse knowledge. Short-term chit-chat, so the thinking goes, does not. So, how good are the best conversational computers?

This question leads us onto the further sense in which Turing is the archetypal AI man. He predicted that by the end of the 20th century computers would pass his Turing Test on about 70% of occasions. In other words, Turing was confident that half a century of progress with computers and a scientific understanding of intelligence would be enough for the scientists to program computers so that they would be able to chat in English just like you or I. Indeed, the essence of the Turing Test is that the computer's language abilities should be indistinguishable from yours or mine.

Just like his scientific contributions, this prediction has set a pattern that endures. The Turing Test pass rate is nowhere near 70%. More than 50 years on, and despite unimaginable advances in computer technology (for Turing we might guess), modern computers are still not close to joining in conversation with us. No computer system has come close to passing the Turing Test. Despite decades of scientific analysis of language and careful programming, and despite hundreds of attempts, the Turing Test pass rate is firmly fixed at 0% with no indications that a lift-off into positive numbers is imminent[7].

A conversational computer remains elusive. Brief banter delivered by computers, as we shall see, is with us here and now, but extended dialogue ranging through diverse topics is still the sole preserve of humanity. Indeed, the average three-year-old toddler can out-talk any computer system.

Ever since Turing framed his test, the field of AI has been rife with predictions of the breakthroughs to be expected "soon"[8]. Over the intervening half century or more, the details of this trend have been relentlessly reconfirmed; all such predictions, just like Turing's, have failed dismally[9].

Yet alongside every new demonstration of an IT system that shows the faintest stirrings of what might become intelligent behaviour, we still get confident predictions that the big breakthrough will be soon — as soon as the next steps are accomplished. Such is the prevalence of this ploy that it has a name: hopeware. It's the anticipated but necessarily hazy successor to the IT system (the software), or the robot[10] being discussed or demonstrated. (Hopeware will get its own chapter once we have surveyed a range of examples).

Why is this? A simple, but hardly enlightening, answer is that we the scientists (enthusiastically over-supported by science writers and IT entrepreneurs) have consistently overestimated the speed of development of the pertinent technologies. This has happened primarily because human intelligence has proved (so far, at least) to be a far more complicated, even mysterious, phenomenon than it appears from a distance.

Is over-enthusiastic prediction also due, more fundamentally, to the nature of intelligence? Can a significant part of the cause be an irrepressible urge deep in the genetic make-up of us all?

Every intelligence is uncontrollably over-eager to "see" a like mind operating elsewhere. This human propensity can overthrow objective judgement even for that somewhat fictional character — the cold and rational scientist.

(iCub photo by Lorenzo Natale, IIT, Italy; reproduced with permission)

This is one of the iCub robots. It looks like a human child, it speaks like one, and it can do things that a human child can do. We meet this "toddler robot" in the 2011 UK-TV Channel 4 series[11], *Brave New World*, where we see that it can chat persuasively, it can learn to recognize objects, and it can understand what it is told to do with these objects. In particular, it can reach out and learn to grasp the objects. Other versions of iCub learn to crawl, learn to shoot arrows at a target, and so on.

The iCub robot also has facial expressions to express its supposed moods. All in all it appears on the way to an Artificial Intelligence well in excess of the Turing Test requirement of textual conversation.

Central to the impressive TV demonstration, the iCub was presented with two objects, named as an octopus and a purple car. It followed instructions "to learn" each object using its eyes, and then "to touch the purple car" which it proceeded to do.

Just after "learning" the octopus, it came up with an off-the-cuff comment that smacked of real intelligence.

Unfortunately, immediately after "learning" the second object, the purple car, it apparently could not resist the urge to broadcast its enthusiasm for this object as well.

One spontaneous, apt and non-functional statement of pleasure gives a fillip to our predisposition to "see" intelligence behind it. But, when apparent spontaneity is immediately and precisely replicated in terms of both the utterance and its context, the bubble of belief bursts — at least for a cynic like me it does.

But this is science, so my prejudices are irrelevant. Experiments could shed some light on this question: are we witnessing the chatter of naïve pleasure that derives from the satisfaction of language-processing competence and knowledge management? Or are we hearing a simple pre-packaged response — one that is triggered by a simple pre-programmed rule? A rule such as:

IF new object X is learned **THEN** say "I like the X very much"

Are we witnessing, in iCub's behaviour, plausible first steps along the road to reproducing, and therefore understanding, intelligence? Or are we caving in to our genetic heritage, and overrating a package of specific, clever tricks devoid of long-term promise?[12] How could we begin to distinguish between these two possibilities?

We might, for example, introduce a sequence of new objects for iCub to "learn" and see what it says. But no such potentially enlightening experiments are ever presented. Additionally, well-directed questions could similarly go some way towards uncovering the details of what's going on in iCub. But iCub's handler is not asked, for example, about the "trigger" for this persuasive utterance. (We'll re-visit this specific question in a later chapter when the necessary background has been established.)

In the course of a scientific discussion, the scientist, Nicky Clayton, espoused her radical theory that birds have developed a degree of intelligence along a separate evolutionary route from humans. The interviewing scientist did ask probing questions, and stated that he found it "difficult to choose the right questions to ask. After all how do you avoid getting answers you are looking for?"[13]

With the TV iCub demonstration we are simply left with its (carefully choreographed?) performance, leaving our eager intelligence detectors unperturbed. "You can almost see its artificial brain learning in front of your very eyes," the TV frontman points out (the same scientist who was alert to the challenge of asking "the right questions" about the intelligence of birds). Sadly, rather like the persistent optical illusions that we'll see in a subsequent chapter, it is true that you can almost see AI at work in iCub, even though there is actually precious little objective evidence for it.

Why do AI presentations, commentaries and media extravaganzas get away with this scientifically vacuous show-and-promise technique time after time? I put the blame squarely on evolution. I suspect that we humans have a deep-seated predisposition to credit a system with "intelligence" just as soon as it reveals any snippet of plausibly intelligent behaviour. Why? Just think about it: for virtually all of our evolutionary history (millions of years) every system that showed any signs of intelligence was intelligent. So this no-brainer decision — is a system that shows a hint of intelligence actually intelligent — became, through continual reinforcement, a fundamental truth, and a solid one until shortly after Turing put his thoughts on paper in 1950.

Here's the root of this irresistible rush to misjudgement, a genetic imperative (and Maxim 5):

If it's like me, it's intelligent.[14]

I first thought that this imperative was: if it behaves like me, it's intelligent. But the persuasiveness of robots with just the loosest similarities to a human being, sometimes just a face, suggests that, although a bit of "intelligent" behaviour is required, some visual similarity is the real clincher.

At the outset of the iCub demonstration we are warned, however, that this "toddler robot... is in the early stages of development". But what stages are they? They appear to be centred on the difficult task of learning motor control of limbs, but the presentation also suggests visual object recognition, knowledge learning and management, and natural-language comprehension — three further and important stages en route to an understanding of intelligence.

The cynic might doubt that this project makes significant inroads into any of these three important aspects of intelligence, and no overt claims are made. They do not need to be; our genetic imperative jumps in to fill the gaps. But if a scientific understanding is sought, what potentially illuminating questions should be asked?

10

Anyone trying to get a grip on the reality of iCub's progress towards AI would, for example, probe the details of its "learning" the octopus and the purple car. What is the knowledge it has "learned", and how was it acquired? With respect to its competence in communicating in English: what is its scope and its limitations? Surely iCub can do more than produce pre-specified responses to half a dozen similarly pre-determined commands. Any competent programmer would need only about 10 minutes to construct the necessary program for iCub to do this. There must be more to iCub's language-processing abilities if they are more than window dressing for its grasping demonstrations, but no one asked and this is all we see it do.

No one ever asks, "What colour is the purple car?"; "Is the purple car a car?" Nor are there requests to "Touch the car." or "Touch the purple object." And so on. Too simple to contemplate you might think — you'd be surprised.

We will see how you, as well as the interested scientist, might probe iCub's specific competences in later chapters. But first we need a little more groundwork followed by some preparatory work on the specifics of language-processing mechanisms and that slippery, but vital, concept: knowledge.

So let's start the foundations with an assumption that is fundamental to most of the relevant science — much cognitive science and virtually all AI. It is known grandly as the Computational Theory of Mind (CTM).

Crudely put, the CTM rests on the presumption that human intelligence is (or can be effectively simulated as) a computer program running on an appropriate computing device.

For those readers of a nervous disposition with regard to mathematics, the next instruction ought to be: look away now, or turn the page quickly. But it is not. In this book we are tackling this phobia head on (but gently). Learning to overcome life's little mental obstacles is to exercise one of the basic functions of intelligence.

Here's the CTM as an equation:

$$MIND = BRAIN + IT\ SYSTEM$$

But if even this small foray into mathematics is causing palpitations, all is not lost; you can try the picture instead which, if successful, might just show that a small mathematical equation is worth a thousand picture elements.

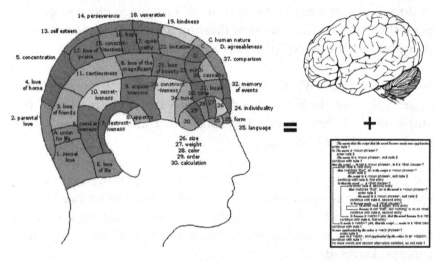

In other words, a basic belief is that intelligence is not inextricably tied to the human brain — the two can be separated into a computer program and a "machine". Our brains are just "meat machines" on which the program, the software or IT system that delivers intelligence, runs. It is, moreover, a machine whose basic design — networks of nerve cells — does not appear to have been fundamentally changed since it first evolved to support the survival of life forms eons before intelligence put in an appearance. Indeed, when shrimps were the vanguard of intelligent life forms and dry land still awaited a first patter of animate limbs, nerve cells were already organizing the crustaceans' behavioural strategies.

Although you might be happy to concede that Martians would not have human brains but could be intelligent nevertheless, the idea that intelligence might be a computer program is a little odd. So let's attempt to reduce this oddity. We'll close in gently on an equational illustration of a couple of homely examples.

The opening quotation for this chapter bears repeating:

"We can't understand the plot of *Downton Abbey* by taking apart the television."

Episodes of any TV soap opera such as *Downton Abbey*, are typically viewed on a television set. But who would consider it anything other than a burst of insanity to probe the innards of their TV during an episode of *Downton Abbey* (with say a voltmeter-ammeter or oscilloscope) in order to better understand what they are seeing? As a consequence of living in this electronics-saturated world you know that the show and its means of conveyance are totally separable. Indeed, TV sets differ widely (e.g., old cathode-ray tubes versus modern plasma screens), and many now watch television programmes on their computer or smart phone. So the details of the viewing device cannot be relevant to an understanding of an episode. This is what we have:

an episode of TV show = + acted script

or

an episode of TV show = + acted script

Whether the viewing experience is delivered by an old TV or a new one is immaterial, although it may affect the quality of that experience. It is the "acted script" that is important, not the details of the device that happens to be conveying the action. Analogously, it is the "script" for intelligence that is important not how the human brain happens to work. This is what the CTM assumption entails.

A further implication of this assumption is that the structure of the machine cannot be determining any essentials of the TV show. They may provide a different picture quality or smoothness of delivery, but nothing germane to the "meaning" of the TV show. What we have is a first instance of the important, undeniable truth — **structure does not provide meaning** (Maxim 4) — which pinpoints the dislocation between understanding at different levels of interpretation of complex phenomena.

A bit closer to our target, here's another example of CTM-like separation. If you are a computer user, you regularly transform the behaviour of your machine by installing various software on it, i.e., computer programs. Two examples are portrayed below:

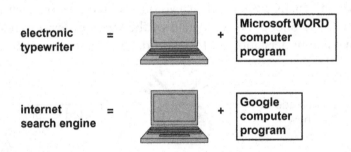

The same computer is made to function as either of two different machines by the software, the program, the IT system that is run on it. Exactly the same configuration of electronics (i.e., hardware machine) can be made to work as any number of entirely different systems. So the structure of the hardware cannot be providing anything fundamental to the behaviour of the system you elect to use — in the above example, an electronic typewriter or an Internet search engine.

It is true that you don't have to install Word, or whatever, every time you want an electronic typewriter; you just click to open and activate it. Just maybe, one day Microsoft Office will contain a "brain" utility — just double click on the ☁ icon, and your laptop becomes intelligent![15]

Similarly, the brain is made to behave as an intelligent machine by the program running on it. This is our basic assumption (but, as we shall see, it is by no means unchallenged).

The plan then is:

1. to find the information-processing principles that underlie intelligence;
2. to use them to construct a computer program; and then
3. to install and run this program on the biggest and fastest computer that's available.

If the resultant computer system exhibits intelligence (perhaps by "merely" passing the Turing Test) then we have some assurance that we have indeed achieved a scientific understanding of intelligence. Some might choke on the CTM presumption, but it captures the working hypothesis, if not fundamental belief, of the vast majority of AI persons and cognitive scientists.

However, unlike the TV drama analogy, we do not have an "acted script" to study. Indeed, it is precisely this "script" that we seek. And moreover, we currently have only one device that runs the "script" we seek — the human brain. This appears to leave us with little alternative other than to explore our single working example with the goal of extracting the script for intelligence. So that's how we'll begin.

Encapsulated in one diagram we have the whole strategy from real intelligence — a working brain — through to an Artificial Intelligence, i.e., a powerful computer system upon which is installed the "intelligence" program.

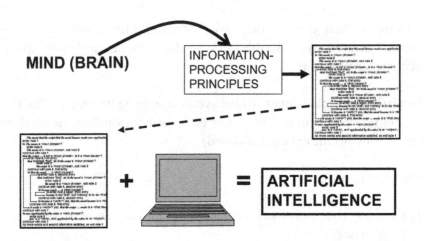

What is being illustrated here (apart from another equation)? First, modern brain scanning technologies allow the neuroscientist and cognitive scientist to sort out the information processing going on inside this "machine" in our head. The subsequent insights into how the brain does the marvellous things we experience it doing, the information-processing principles extracted, are written into a computer program. This IT system is then installed on the best computer system available. This computer will then behave intelligently. That's the overall plan.

By these means, an Artificial Intelligence would have been created, and moreover the scientists who wrote the successful program will have a secure basis for their understanding of intelligence. This is the further assumption that AI scientists and many cognitive science researchers rely on:

> If this AI program can be constructed, then science has understood intelligence. It will know how intelligence works.[16]

What's the alternative to the CTM assumption? That brains and basic intelligence are *not separable*[17]. In which case, to understand intelligence the scientists would have to build a simulation of the pertinent aspects of the human brain. Not everything about the brain is relevant because it does much more than "think". But what is pertinent to its "thinking" function, and what aspects are not? Until we understand more about how "thinking" works, we cannot hope to work out which aspects of brain function are irrelevant. There is a nasty circularity here, and so science wisely avoids this avenue of approach. In the next chapter, we will see this "salience" problem arise in the details of the scientific investigation.

The importance of this stumbling block is signified by making it one of the ten maxims — number 3: the vicious circularity of **understanding based on a single example**.

One of the latest predictions, solidly within the time-honoured AI tradition, is the advent of the so-called Singularity which we are to expect within the first few decades of this new millennium. The Singularity prediction is most comprehensively promoted by Ray Kurzweil, a very successful US AI entrepreneur, in the 600 or so pages of his 2005 book *The Singularity is Near*.[18] The book's message is endorsed by luminaries such as Bill Gates, a man whose take on computer systems must be quite solid, one would think.[19] This idea has been extensively promoted in the USA and in the UK.

The Singularity is thus no straw man, but what is it exactly? According to Kurzweil, the Singularity, "the destiny of the human-machine civilization" (p. 5), has as its most important implication that "our technology will match and then vastly exceed... what we regard as the best of human traits" (p. 9). He subsequently asks, "What are the consequences of this event? When greater-than-human intelligence drives progress ..." (p. 21).

Rather than deal with the full-blown Singularity and its mind-blowing consequences, I shall focus on its much more modest precursor: a computer system that simulates most of the intelligence exhibited by virtually every specimen of humanity. So, a system that could pass the Turing Test, but not one that necessarily displays human emotions, or even consciousness. These latter properties might just emerge, but all

17

that I want to concentrate on is the vastly simpler task of understanding basic intelligence sufficiently well to program a computer so that it can listen, speak, see, reason, etc., just like you or I. If it displays emotions, becomes infatuated with an adjacent slinky filing cabinet, or insists that it's conscious, that would be very interesting, but not my central concern.

This is a much more modest interpretation of the Singularity proposal, perhaps just a first step along Kurzweil's grand parade, but that's quite enough to think about tackling. Unsurprisingly, we are given a relevant prediction: "We will have effective software models of human intelligence by the mid-2020s." (p. 25)

As a long-term AI researcher, it is only fitting that I join in fully and offer my prediction. I endorse the above prediction in its entirety once the word "not" is inserted between "will" and "have". There, I've stuck my neck out, and if history is not to be turned on its head, my prediction will also fail. But as these two predictions are mutually inconsistent, they can't both fail, can they? Actually, they probably can. The word "effective" may introduce ample wriggle room for arguing the reality of software models in either direction, come the end of the 2020s. If this were indeed the case in 2030 then Singularity advocates would have to concede that progress had not been quite as fast and furious as expected, and I would have to concede surprise at the speed of progress in AI.

So let's quit the fundamentally unsatisfactory "oh yes it will" countered by "oh no it won't" debate on the future of AI as a science. We must examine the details of the scientific research and see what we do know about how intelligence works.

To begin with, we can follow the work plan that enthusiasts expect will deliver the Singularity, and we'll see how well it stands up to scrutiny. Singularity advocates want to start by understanding the brain, and that does appear to be a sound and indisputable basis for developing a scientific understanding of intelligence. Although brain science is not an area of expertise to which many readers can lay claim, it is just one place to start.

So-called "reverse engineering" is the primary idea proposed: construct an intelligent system from information gleaned from detailed investigation of the human brain. It is this plan to progress from brain study to the "intelligence" program that is illustrated in the last diagram. This venture is solidly supported by explosive growth in all the attendant technologies: brain scanning technologies to deliver the evidence from which the principles underlying intelligence will be abstracted, and computational technologies to provide the power and capacity for implementing the principles extracted.

At some risk of blunting the keen edge of expectation, I'll very briefly say why this strategy of reverse engineering is doomed to failure. In a text-bite that encapsulates the central obstacle, which we've already revealed in the context of home-computer usage and TV shows: structure is not sufficient to provide meaning. This is an echo of one of the four clinchers[20] offered by philosopher John Searle. The four together underpin his claim for the hopelessness of the AI endeavour. For me, this undeniable truth — **structure does not provide meaning** — is not an endorsement of his impossibility argument with respect to AI. It is a borrowed observation specific to reverse-engineering; the effect is that brain-scan evidence can *never* deliver meaning in terms of intelligent behaviour. A seemingly bottomless abyss separates the measurable structure and operations of a working brain from the "meaning" of what this brain is doing in terms of intelligent thinking.

Consider one more homely example: with CCTV and other spying technology running riot by 2020 (even nanobots crawling throughout your car), all the details of, say, a drive from Exeter, in the southwest of the UK, to London will never reveal the purpose of the journey. Full details of the route and even the principles and patterns of your driving behaviour could reveal a lot about you as a driver but nothing about *why* you are making the journey. The fact that you choose to drive along the A303 will be noted, but nothing about your desire to check on the continued existence of Stonehenge, as the motivation for the route choice, is likely to emerge. Much of the remainder of this book is devoted to fleshing out this fundamental point, although a number of substantial subsidiary further difficulties arise and will merit attention.

Let me repeat, this is not to deny the possibility of ever building an artificially intelligent system. It is to assert that the necessary information cannot be gained just by monitoring the structure and operational details of one working example: the human brain.

The term Singularity, seemingly adopted to reflect the runaway prospects of an ultra-intelligent machine, is for most of us associated with the strange notion of an astrophysical black hole from which nothing, not even light, escapes. As we shall see, the reverse-engineering proposal smacks of a black-hole singularity: the brain-scan data disappears in, and nothing vaguely interpretable as the principles underlying intelligence ever emerges.

To present the positive side, not because I do not see anyone making a TV programme, *The Singularity is Nowhere Near*, but because I do see the glass of AI as approaching half full. Human intelligence has revealed itself to be surprisingly enigmatic, which is fascinating in itself, but associated sciences are making progress. I see science as slowly but steadily adding to the contents of the glass, even suddenly filling it to overflow one day perhaps. But I do not see this happening primarily via reverse engineering, and I do not see it happening in the next few decades unless some startling breakthroughs occur, and they just might (but I'd strongly advise against any holding of breath while waiting).

At arm's length we can kid ourselves that we are beginning to understand the organization of cognition (so-called short-term and long-term memory aspects, for example). However, as we shall see, the more closely we grapple with the essentials of human intelligence — like learning and forgetting, or commonsense reasoning versus specialized knowledge — the more our theories about what's going on are revealed as inept.

The difficulties are three-pronged:

1. The difficult nature of the phenomenon itself, intelligence;
2. The fundamental difficulties with the experimental and modelling medium, Information Technology; and
3. Difficulties with the investigating system, humanity — difficulties from which scientists are not exempt.

These three sources of unhelpful characteristics conspire to pervert the course of scientific practice, making an already huge scientific challenge even trickier. Ten maxims that crystallise the essence of these difficulties are explained and justified in a glossary, and will be called upon throughout. We've already made a first call on three of them which were:

3. A vicious circularity: **understanding based on a single example.**
4. An undeniable truth: **structure does not provide meaning.**
5. A genetic imperative: **if it's like me, it's intelligent.**

It is time to see the basic workings of science and at the same time uncover the sticking points when science attempts to lay bare the principles that underlie intelligent behaviour by studying the workings of our brains.

Endnotes

[1] Attributed to psychologist Rufus May by Jim Al-Khalili (although he actually said *EastEnders* not *Downton Abbey*) in a discussion with Robin Murray about treatments of schizophrenia on UK Radio 4's *The Life Scientific* broadcast 7th February 2012. The issue here was one of treatment via "talking therapies" (based on social sciences) or medication to alter brain behaviour (based on neuroscience).

21

[2] Alan Turing's musings on AI are to be found originally in the academic journal, *Mind*, October 1950, vol. 59, pages 433–460, but reprinted many times and perhaps more accessible in, for example, pages 11–35, *Computers and Thought* edited by E. A. Feigenbaum & J. Feldman and (McGraw-Hill, 1963).

[3] Many have taken a swipe at Turing's Imitation-Game proposal. Open almost any AI book and you'll find a discussion of it. In particular, Mark Halpern's 1987 "Turing's test and the ideology of artificial intelligence" published in the *AI Review* (vol 1, no. 2, pages 79–93) offers a real demolition job. Others, equally well informed, deny that Turing's proposal was intended as a test of intelligence — see, for example, Aaron Sloman's argument on his website at www.cs.bham.ac.uk/research/projects/cogaff/misc/turing-test.html.

[4] A recent, non-technical, report on the Loebner version (see later note) of the Turing contest is given by one of the humans whose job was to "be human" (i.e. system A) for comparison with the computer systems competing — see *The Guardian Weekend*, 30th April 2011, pages 17–20 by Brian Christian who has also written a book, *The Most Human Human: a defence of humanity in the age of the computer* (Viking Books, 2011).

[5] This also introduces a major point of contention with Turing's proposal: how long is "an extended dialogue"? If the interrogator fails to spot the computer, it can always be argued that the dialogue was too short. Currently, this is not a real issue because the possibility of intelligent dialogue with a computer typically evaporates in the first few exchanges.

[6] As stated in the F. E. Sutcliffe translation of *Descartes Discourse on Method and Other Writings* (Penguin, 1963) pages 73–74.

[7] Writing in the second decade of the new millennium, the Turing Test remains to be passed even once. (See subsequent note on Loebner Prize competitions.)

[8] See, for example, *Reasons to be cheerful in 2012: it's the year we'll start talking to robots*, in *The Guardian*, 20th January 2012 by Jon Ronson, "Finally, we're all going to be able to have a proper conversation with a robot. Jon Ronson, for one, can't wait... For human-*computer* conversations, 2012 is going to be a great year. We might actually start having proper, fluent exchanges with robots, *chatting* away as if we're the same species..." December 2012 has come and gone, re-check this hope at the end of 2014, or 2015, or... who knows?

[9] "In 1990 Hugh Loebner agreed with The Cambridge Center for Behavioral Studies to underwrite a contest designed to implement the Turing Test. Dr. Loebner pledged a Grand Prize of $100,000 and a Gold Medal for the first computer whose responses were indistinguishable from a human's. Such a computer can be said "to think". Each year an annual prize of $2,000 and a bronze medal is awarded to the **most** human-like computer. The winner of the annual contest is the best entry relative to other entries that year, irrespective of how good it is in an absolute sense." Statement on the Loebner Prize website (accessed 4/1/2011).

    To this day, the Gold Medal and the $100,000 cash remain in the care of Dr. Loebner. Despite many contests, there has not been the slightest hint that he has been tempted to dust off the Gold Medal and alert his bank manager to the possibility of a withdrawal beyond his cash card limit.

[10] These robots, despite some impressive engineering, are primarily controlled by computer programs to provide their ability to listen and to speak as well as their hand and eye movements. So in all examples the essential intelligence is provided by a computer program, a piece of software.

[11] The first programme in the series *Brave New World with Stephen Hawking* broadcast on UK television by Channel 4 on 17th October 2011.

[12] Hubert Dreyfus, an early well-publicized critic of AI who gets considerable exposure in later chapters, once likened what I've called the limited clever tricks of AI demonstrations to climbing a tree as the first steps towards getting a man on the Moon. If you climb a tree you're certainly closer to the Moon, but improvements in tree-climbing, however amazing, are never going to get you anywhere near a landing on the Moon.

[13] *The Life Scientific* broadcast in the UK on BBC Radio 4 on 22nd November 2011 at 9:30 am. Nicky Clayton talks to Jim Al-Khalili about her work on the intelligence of birds — two scientists discussing a theory.

[14] It has been pointed out to me by Antony Galton that humanity's history of slavery, racial discrimination and apartheid regimes seems to argue against this supposed predisposition I offer. This history can be interpreted as a human tendency to reject the humanity (and hence intelligence) of others who differ only slightly, e.g., different skin tone. Perhaps these ugly phenomena are political and social amplifications of tribalism, another genetic predisposition, and one that causes us to favour those most similar to us (selfish gene territory?). Can we account for the AI-triggered tendency to "see" intelligence at work because the "tribal" counter-weight is not yet operative when the observed systems are clearly not (yet) competitors?

[15] Although I'd advise strongly against delaying purchase of your next computer system until the "intelligence" utility is available, it's an intriguing idea. For example, would your "intelligent" laptop immediately dump your Microsoft systems and start searching the Internet for something more secure?

[16] Even at this early stage, we note a fundamental clash between this operating principle and one of our maxims, i.e., no significant computer program is completely understood. The possession of an IT system that behaves intelligently will always facilitate argument over the roles and significance of certain details, but the scientists will have an explicit computer program to argue about, test and probe. This would be a monumental advance on our current state of scientific ignorance.

[17] Many scientists argue for the non-separability option. In his "emergent theory" of mind Victor Johnston, for example, says that "sugar is not sweet, nor toxins bitter, for all such pleasant or unpleasant feelings have evolved as discriminations between a diverse range of potential threats or benefits to our [biological] survival" and so "events [in the world outside our heads] have no meaning in and of themselves, they just are. It is the emergent properties of neural pathways, shaped by the life and death

23

of our ancestors, that provide meaning to these otherwise insignificant physical-chemical events", from his 2003 paper "The Origin and Function of Pleasure" in the journal *Cognition & Emotion*, 17, pages 167–179. In other words, crucial features of human intelligence are dependent upon aspects of our evolved neural organization. Similarly, intelligent Martians would inevitably have a different set of feelings, evoked by the environmental threats and benefits required for survival on Mars, and these would be a consequence of an evolved Martian brain organization. The key issue is then: is there a common and brain-independent core of principles that we might justifiably call "basic intelligence", or is "unfeeling intelligence" necessarily an oxymoron? An earlier, but more accessible and comprehensive, explanation of this objection to the CTM is provided by Johnston in his book *Why We Feel* (Helix, 1999).

[18] The term Singularity as ultra AI was perhaps first promoted by Vernor Vinge (a US academic) in a 1993 lecture in which he predicted its arrival in the next thirty years, or (after acknowledging AI's history of time-ambiguous predictions) stated it as: "I'll be surprised if this event occurs before 2005 or after 2030." A 2013 documentary film by Doug Wolens, *The Singularity, we will survive our technology* is accessible via thesingularityfilm.com. *IEEE Spectrum*'s (a respected US technical magazine) June 2008 issue has a 50-page special report on the Singularity. It is in part an attempt to see how the Singularity measures up against real-world science and engineering. Or as stated on the cover: "The Rapture of the Geeks: Separating science from fiction in the technological singularity." MIRI, the Machine Intelligence Research Institute based in Berkeley, California, gives considerable coverage to the Singularity issue including publication of *Singularity Hypotheses: A Scientific and Philosophical Assessment* (Springer, 2013) which "offers authoritative essays and critical commentaries on central questions relating to accelerating technological progress and the notion of technological singularity, focusing on conjectures about the intelligence explosion, transhumanism, and whole brain emulation."

In the UK, the *Horizon* programme broadcast by the BBC on 24th October 2009, trotted out the simplistic thesis that computers will very soon be as powerful as human brains, seemingly oblivious to the multi-dimensional vagaries of such an assertion, e.g., calculating versus commonsense reasoning. A few intriguing projects (such as attempts to rebuild a mouse's brain piece by piece, and monkeys that learned to control levers using brain signals) were briefly visited but with no real attempt to explain how these projects were leading to the Singularity in the near future.

[19] Ray Kurzweil's *The Singularity is Near* was published by Viking, Penguin in the USA and by Duckworth in the UK both in 2005. Bill Gates is quoted (among many others, although leading computer scientists as well as AI scientists are curiously almost absent) as saying, "Ray Kurzweil is the best person I know at predicting the future of artificial intelligence." Bill obviously needs to get out less.

To mention "the exception that proves the rule", it should be noted that Kurzweil did predict that chess-playing computers would be better than humans by 1998, and IBM's Deep Blue system did indeed defeat the World Champion, Garry Kasparov in

24

1997. But see the December 2010 edition of the *IEEE Spectrum* technical journal in which J. Rennie wrote an article entitled *Ray Kurzweil's Slippery Futurism*. In it he made the case that Kurzweil's "stunning prophecies have earned him a reputation as a tech visionary, but many of them don't look so good on close inspection". After examining and explaining the absence of substance in a variety of predictions, Rennie writes, "Most of his [i.e., Kurzweil's] predictions come with so many loopholes that they border on the unfalsifiable. Yet he continues to be taken seriously".

[20] "Syntax is not sufficient for semantics," wrote Searle, "That proposition is a conceptual truth. It just articulates our distinction between the notion of what is purely formal and what has content." This is from John Searle's famous anti-AI book *Minds, Brains and Science* (Harvard University Press, 1984), p. 39. My use of Searle's textbite is not an endorsement of his impossibility argument for AI — I believe AI can and will happen. I only echo his syntax without adopting his full semantics.

෬ Chapter 2 ෬

Scanning for Gold

"Our ability to reverse engineer the brain — to see inside, model it, and simulate its regions — is growing exponentially. We will ultimately understand the principles of operation underlying the full range of our own thinking, knowledge that will provide us with powerful procedures for developing the software of intelligent machines." (p. 144)

"One simple statement of the strong AI scenario is that we learn the principles of operation of human intelligence from reverse engineering of all the brain's regions, and we will apply these principles to the brain-capable computing platforms that will exist in the 2020s." (p. 293)

both Ray Kurzweil, 2005[1]

The main goals of this chapter are two-fold: to explore the problems that arise when the scientist attempts to extract the principles underlying intelligence by studying a working brain; and, at the same time, to illustrate the fundamentals of scientific enquiry. The problems that arise point to the importance of *feature salience*. They also lead to a better understanding of the gap between structure and meaning (in effect, Maxim 4). In particular, we will see the necessity to introduce *levels* of understanding.

Before we delve into how brains work, we can take a first dip into the essential workings of scientific investigation by considering the card game Eleusis.

In Eleusis the deck is divided among the dealer and the players. The dealer thinks up a rule that defines a card sequence, and he or she provides an example, some evidence, of the rule.

The dealer's evidence might be:

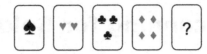

The question to be answered by the other players is: what rule is being illustrated by the evidence of these first four cards? Each player in turn looks at the evidence presented and guesses what this rule might be. They are looking for a pattern in the evidence, the pattern that is in the dealer's mind. Each player forms a hypothesis to define the underlying pattern they "see". In this case, the hypothesis of the first player might be:

HYPC1: *each card is one greater in face value than the preceding card.*

The player can then test this hypothesis by performing an experiment. He or she places a fifth card down. Then the dealer will deliver the result of the experiment as either *correct* (if the fifth card follows the rule) or *incorrect* (if the fifth card does not follow the rule).

On the basis of HYPC1, what card should our player use as a test of the hypothesis? What experiment should be performed? The obvious choice would be one of the four cards with a face value of five. Does it matter which one? Not according to our hypothesis, but let's see. Suppose this player holds the five of spades and plays it as the experimental probe.

The result of this experiment is given by the dealer as: *correct.* Does this mean the hypothesized rule, HYPC1, is correct? No, not necessarily. In fact if the player had played the five of diamonds instead, the outcome of this experiment would be: *incorrect.* So HYPC1 cannot be the dealer's rule.

Imagine yourself as the player here. Which outcome from your experiment would you prefer? A no-brainer. It is much more satisfying, and seemingly productive, to be declared *correct* than *incorrect* — an experimental success is preferable to failure.

But notice that a *correct* experiment only tells the scientist that her hypothesised rule **might be** correct. It is the *incorrect* experiment that gives hard information: the hypothesised rule **is** incorrect. This mismatch of personal satisfaction with value of outcome is, I will argue, an important element of the psychological tangle that obscures the scientific quest to understand intelligence. Remember how impressed we all were that iCub could successfully touch the purple car when so requested?

To continue, we know that the hypothesised rule, HYPC1, cannot be the underlying rule because the five of diamonds follows the HYPC1 rule and yet is *incorrect*. In the light of the evidence from these two experiments, we might modify HYPC1:

HYPC2: *each card is one greater in face value than the preceding card and from a different suit.*

Both experimental outcomes are consistent with HYPC2, but that does not guarantee that this hypothesised rule is the one that has given rise to the evidence available. It just might be.

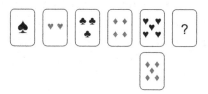

So further experiments can be performed on this larger quantity of evidence that is now available (the out of line card indicates that it was incorrect at that point in the sequence).

We have in Eleusis a crude analogy with the basics of scientific practice. The dealer's rule is like a law of nature. The first four cards are evidence consistent with the natural law, the underlying principle. The players are scientists experimenting to test their hypotheses in an effort to discover the law that is generating the evidence. Each experiment adds to the evidence available.

The basic framework of scientific enquiry can be summed up:

- consider the evidence;
- devise a hypothesis — a proposal for the underlying law — consistent with the available evidence;
- test the hypothesis by performing experiments;
- assess the experimental result — if the outcome supports the hypothesis, test further; if the experimental outcome does not support the hypothesis, develop a new hypothesis from the wider range of evidence now available[2].

In particular, notice that should a player now place, say, a six of hearts next, which the dealer then declares *correct*, the player would be foolish to cease the experiments convinced that HYPC2 is also correct. All that this shows it that HYPC2 might be correct. What the scientist cannot do is try one experiment and conclude that his or her hypothesis is correct when the experimental outcome is in accordance with that hypothesis.

Suppose instead of the six of hearts, the next experiment tries the six of clubs and the dealer declares this *incorrect*? The player proposing HYPC2 would know that it is not the sought-after sequence rule[3]. The theory encapsulated by HYPC2 has been disproved; it has been refuted.

For the scientist, failed experiments are more informative than the successful ones: a failure tells you that your hypothesis is definitely wrong, whereas a success only tells you that it might be right.

Hence this important "sad" fact: **failure drives science forward** (Maxim 7). This observation is an inexorable consequence of the logical conundrum at the heart of science: there are no proofs of correctness of theories[4], only refutations. And so, **we cannot prove, but we can disprove** (Maxim 6). This inescapable logic grates on the human urge to understand and know. So much so that non-scientists almost never accept it[5], and many scientists just try not to think about it as they go about their research. Scientific journals are always reluctant to publish "failed" experiments even though they can be more informative than those with successful outcomes[6].

Here's a summary statement of experimental science. It combines the logical conundrum with the human urge to explore and understand:

Successful experiments are most pleasing, but unsuccessful ones are most informative.

Couple this summing-up with the earlier genetic imperative that operates when intelligence is the topic being studied, and we begin to see the basis for the ever-gushing fount of over-optimistic prediction in AI.

In our particular context, the card sequence rule is analogous to a rule of operation of some aspect of intelligent behaviour. It might, for example, be the rule that underlies human learning. The cognitive scientist studies examples of this learning behaviour and frames a hypothesis that can explain the evidence. The next step is to devise experiments that test the hypothesis, and so on.

We will now transfer this experimental framework to the evidence to be found in a working human brain. The scientist can use brain-scan evidence to discover the rules that govern the behaviour of a "thinking" brain, and so perhaps elucidate the principles underlying intelligence. This is precisely the "reverse engineering" strategy introduced towards the end of the previous chapter.

Delving into brain structures and their workings does move our enquiry beyond the bounds of everyday expertise. So, initially at least, most readers will be denied the uncommon luxury of questioning the

31

science from personal experience, but that will not last too long. Right from the outset, you will see that the issues arising are well within the compass of critical consideration that non-neuroscientists like you and I both possess. Here's the overall picture:

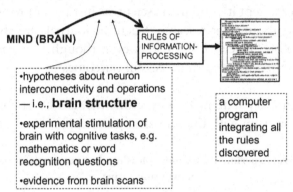

It is brain-scanning technologies that provide the scientists with the evidence needed to devise and test hypotheses about how the brain works. It is time to take a stroll along this road to the future in order to get a feel for the details of its route as well as check for potholes and the like.

First, let's view the "machine" of interest, the human brain, and then the experimental evidence it offers the eager scientist:

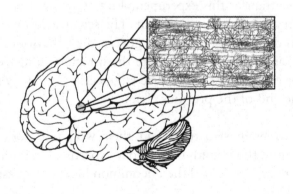

As you can see, the human cortex looks to be primarily a tangled mass of highly-interconnected nerve cells, neurons[7]. Closer inspection, however, does reveal some anatomically individual substructures for the scientists to study. There are some well-defined components of intelligent behaviour that are associated with distinct structures in the brain (such as columnar structuring and discernable layers of neurons in the cortex). Or the other way around: there are marked substructures within our brains, and some of these appear to be more or less dedicated to well-defined aspects of our cognitive behaviour. It is one such brain component — an anatomical brain module — that we shall focus on.

As a subject for mid-twenty-first century scanning technology we'll look closely at one tiny brain structure, one that accounts for a small part of a brain's intelligent behaviour. The remarkably small brain module I've chosen is composed of only nine neurons — it accepts five input signals, processes them by means of three neurons, and generates a single output signal to pass on to other brain components. Internal to this small brain module there are just 18 synapses to be scanned in order to monitor the internal information flow[8].

Crucially, apart from its five input channels and its single output, this brain component has no other connections with the rest of the brain. Its processing of the five incoming signals is thus isolated from all other brain activity. It is said to be an *informationally encapsulated* module. This is another way of saying that a joint in the brain has been found, allowing scientist to "carve out" this component and experiment with it uncluttered by whatever else the rest of the brain may be doing.

"In order to reverse engineer the brain, we only need to scan the connections in a region sufficiently to understand their basic pattern. We do not need to capture every single connection." And although the necessary technological capabilities "exist at least in an early stage today … we can conservatively anticipate the requisite nanobot technology … during the 2020s" (both p. 166).

"Nanobots" are envisaged as microscopically tiny autonomous "machines" that could be used (in our context) to operate inside a working brain and measure the strengths of signals transmitted between nerve cells, or neurons. Presumably they would report (or perhaps store) the signal-strength evidence for the scientists to hypothesise about, and to manipulate by means of experiments. The current hard reality of brain-scan technologies is the subject of the following chapter.

Thus my proposed scanning, which may be a bit over the top in terms of level of detail necessary and ahead of the curve of current brain-scan technology, is soon to be swamped by the exponentially growing waves of technological reality. To err on the side of too much detail is surely better than that of too little in our quest "to replicate the salient functionality of biological information processing" (p. 443). Too much detail necessitates only the neglect of information whereas too little requires the generation of more information.

It's as if the Eleusis player included coffee stains and bent corners as features of the cards to be included in hypothesised rules. Repeated experiments should reveal the irrelevance of these features, but discovering that card colour is a salient feature when it has been overlooked (as in the chapter-opening example) can be difficult.

In general, the reverse engineer would have to beware of masses of inessential details masking the salient patterns sought, but in our little study surely this cannot be a problem. We will, however, revisit this question later when we have some examples of potential features of brain behaviour at hand.

Here is the small brain module that we've been fortunate to locate in a region of the brain that is always active when the subject's brain is stimulated by a specific geometry reasoning task:

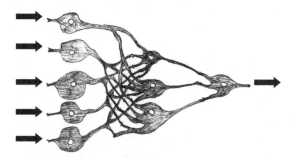

Each of its nine neurons has a cell body containing a nucleus, input connections from other neurons (dendrites), and an output connection (an axon) that may divide to give input to one or more other neurons. There is a layered structure in this module: five "input" neurons (each of which has one input connection from somewhere else in the brain); three "internal" or "hidden" neurons that each have five input connections, one from each of the input neurons, and one output connection; and just one "output" neuron that receives an input from each of the three hidden neurons and itself has a connection out to other brain structures.

Our goal is to reverse engineer just the cognitive aspects of the human brain and not to reproduce a brain in all its physiological detail. So we will need to extract just the essentials. We can neglect the biochemical and molecular details to focus on "the human brain's principles of operation" (p. 25).

Consequently, we are interested in the details of the strength and frequency of the signals that pass along the inter-neuron connections, or links. We are free to neglect the chemistry of neuro-transmitters and receptors that accomplish this signal transfer in the actual wet-ware version of our small brain module. Such molecular details of brain function cannot be necessary features of the "machine", can they? They are akin to the specific red and black pigments used to print hearts or spades on the cards. Some pigment choice is necessary but which ones are actually used is immaterial to discovering the dealer's rule.

The chemical make-up of the particular "processors" that happen to constitute a human brain is deemed to be incidental, just as you can upgrade your computer by changing its electronic innards (e.g., switching to a new and faster processor) with no effect on what your computer can do, and cannot do. It just makes it faster — hopefully. So, the reverse engineer can ignore the chemical details of the brain's "processors".

Armed with such assumptions, our brain-scanning only needs to focus on what, when and how one neuron sends information to another.

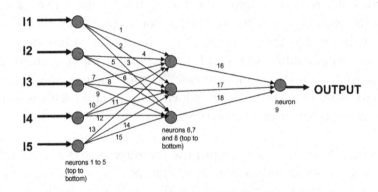

So, the molecular details of synaptic connections and information transmission have been omitted from the structure that we wish to understand. The salient (we hope) bare bones of the brain module are the nine neurons and their 18 inter-connections. With the former as circles and the connections as straight lines, our brain module has been sanitised and clarified. From this point on, the messy biological and biochemical workings of a real brain are banished as unnecessary clutter. This is our brain module; it's time to discover the information processing principles that it uses.

Temporarily leaping ahead of current brain-scan technologies, we'll assume that somehow sophisticated, non-disruptive, single-cell monitoring is available for all nine neurons. This means we can collect detailed evidence about how this part of the brain works, and so develop hypotheses to articulate the rules, or processing principles, that it uses.

The evidence will consist of measurements of the magnitude of all the signals that pass through this network as it goes about its business of contributing to the overall collection of behaviours that we call "intelligence". When primed with five input signals we can collect the evidence of how it processes these signals and so generates an output signal. Our experiments will be variations in the five input signals designed to test the hypotheses we develop.

Pushing on with our experiments, we first note the parallelism in the information processing of our neural-net module — neurons 1 through 5 all get information at the same time (say T1); a short time later (T2) they each send information along their axons (links 1 through 15); at time T3 neurons 6, 7 and 8 all process the information flowing in from their dendrites; at T4 these three neurons each send information out along their axons (links 16, 17 and 18); at T5 neuron 9 gets information in; and at T6 neuron 9 sends out information along its axon. During this information processing period (T1 to T6) no other activity occurs in this network — no new inputs and no self-organisation (i.e., no learning takes place).

Our scan has clearly revealed a pattern of information processing: a parallel flow of information into neurons 1 through 5, then through neurons 6, 7 and 8, to finally reach neuron 9 and out of our little module. Our putative scanners do better than this. In addition to timing information they report the strength of the signals as they pass through this network. Below, you can see the evidence collected from one experiment: the brain module is activated with values for the five signals called I1, I2, I3, I4 and I5.

T1	T2	link	T3	neuron	T4	link	T5	neuron	T6
I1 is	0.2345	1							
3.1478	0.2005	2							
	0.6645	3							
I2 is	19.346	4							
0.1399	15.562	5	89.864	6	0.9870	16			
	1.4404	6							
I3 is	26.878	7							
1.0012	12.234	8	41.567	7	1.9894	17	3.0007	9	0.0023
	65.546	9							
I4 is	9.9982	10							
4.9989	12.342	11	97.623	8	0.0041	18			
	0.0234	12							
I5 is	33.456	13							
3.4451	1.3432	14							
	30.001	15							

This is a big bunch of numbers from just one experiment on only nine neurons. The eyes of every reader have a perfect right to glaze over at this point. Just bear with me and I'll pull out the patterns of information processing revealed and frame them as hypotheses. But do spare a thought for the reverse engineer who must find these patterns in the sea of numbers generated by brain structures composed of hundreds or even thousands of neurons. Too horrible to contemplate, so let's not. We'll just whiz on while I do the job on the evidence of these particular numbers.

So what further information-processing patterns can we now find in our numbers? Well, we can see that neurons 6, 7 and 8 (T3 boxes) as well as neuron 9 (T5 box) appear simply to add their inputs together. For example, the five signals in the T2 column that feed into neuron 6 (the links 1, 4, 7, 10 and 13) add up to a total of 89.9127 and the scan of neuron 6 reveals an internal signal strength of 89.864. A similar examination of the signals into, and out of, neurons 7, 8 and 9 provides further support for this conjecture.

So we have a first hypothesis about how the brain's nerve cells work. This is it:

HYP1: *that a neuron adds together all the input signals it receives.*

We can test this hypothesis (as well as the earlier observation about parallelism) by arranging for repeated sets of five input signals to be fed into this brain module, and collecting all the resultant internal signals as evidence that may or may not support our hypotheses.

You'll have to trust me on this (which is infinitely preferable to leafing through dozens of pages of tabulated numbers) when I tell you that we do find that these repeated experiments are all supportive of HYP1 and also always exhibit the parallelism first observed.

In addition, running with hypothesis HYP1 further suggests that our scanned numbers are too precise (spurious precision[9] generated in the engineering of our scanner or by the natural imprecision of biological mechanisms). If we collapse our recorded values down to, say, one decimal place[10] then our summing hypothesis, HYP1, looks firmer: the inputs to neuron 6 add up to 89.9 and the internal signal strength is recorded as 89.9, for example.

This gives us a second hypothesis, another step towards extraction of the full operational principles:

HYP2: *that the scan data is only accurate to one digit after the decimal point.*

If true, this means, for example, that the numbers 12.2 and 12.3 are considered as different, whereas 12.33 and 12.31 are essentially the same, both will be rounded to 12.3. Here's the "rounded" version of our data. It also contains a few explicit connections that are intended to clarify the information processing of our brain module, in terms of current hypotheses and mysteries (denoted by a simple "?").

T1	T2	link	T3 neuron	T4 link	T5 neuron	T6
I1 is	0.2	1				
3.1	0.2	2				
	0.7	3				
I2 is	19.3	4	add?	?		
0.1	15.6	5	89.9 6	1.0 16		
	1.4	6				
I3 is	26.9	7		?	add? ?	
1.0	12.2	8	41.6 7	2.0 17	3.0 9	0.0
	65.5	9				
I4 is	10.0	10		?		
5.0	12.3	11	97.6 8	0.0 18		
	0.0	12				
I5 is	33.5	13				
3.4	1.3	14				
	30.0	15				

So, we hypothesise that there is some addition of signals being done within the brain, but there is clearly more to the story than that. Now we need to concentrate on the unqualified question marks — they indicate points where hypotheses about the information processing principles still need to be teased out of our numbers.

We can see that the outputs of neurons 6, 7, 8 and 9 have become 1.0, 2.0, 0.0 and 0.0, respectively. So we appear to be dealing with processing units, the neurons that generate whole numbers, perhaps essentially digital, signals, such as 1 and 0.

It is known from the wealth of other neuron studies that these processing elements do tend to operate as two-state devices, either they produce an output signal when stimulated, say a 1, or they do nothing, say 0. So the initial evidence supports the following hypothesis about neuron output signals:

HYP3: *that neurons operate as binary devices — they simply output either the signal 1 or they output nothing, a 0 signal.*

As a consequence of the above hypothesis, we might further hypothesise:

HYP4: *that the 1 or 0 output signal from each neuron is determined by comparing the internal sum with a threshold value; if sum is greater than the threshold then output 1, otherwise output nothing.*

Once more, with further experiments, we find that HYP3 is repeatedly supported. So we build on it and conjecture that the "strength" of link 7 (perhaps synaptic facilitation in biochemical reality) is twice that of link 6, and so on. The effect of these link "strengths" is to amplify the signal transmitted by, say, multiplying each neuron's output by the "strength" of the link that transmits it.

This gives us a hypothesis about the information-processing from one neuron to another. It is:

HYP5: *that a neuron's output signal is multiplied by the "strength" of the link it traverses.*

With the evidence collected from many experiments, our various hypotheses can be further corroborated or eliminated. As it happens, all our hypotheses receive further support, and no contradictory evidence emerges. So we adopt all our hypotheses without further ado.

Shazam! We have a theory of how this brain module might operate. We have a plausible set of operating principles for this particular brain module. We have an explanation of its information processing.

In summary: it seems that the neurons add all their input signals together, so neuron 6 adds together its 5 input signals, (link1 + link4 + link7 + link10 + link13), and this total is "thresholded"[11] to be an output signal of 1 if the total is greater than some threshold value, and 0 (essentially no response) otherwise. Let's denote this summation of these five input signals as $\Sigma(1,4,7,10,13)$. (We confidently expect further experiments to shed light on the value of the thresholds, and for neuron 6 I shall call its threshold *thresh6*, and similarly for neurons 7, 8 and 9.)

So now we know the essential information-processing principles of this brain module. For example, neuron 6 works as follows:

> **if** the result of adding together the five input signals to neuron 6 is greater than the value of *thresh6*,
>> **then** neuron 6 sends an output signal,
>> **otherwise** it does nothing.

And with our bit of mathematical notation introduced above (as well as the symbol ">" for "greater than"), we can restate this information-processing principle a little more succinctly as:

> **if** $\Sigma(1,4,7,10,13)$
>> > *thresh6*
>> **then** signal 1
>> **otherwise** signal 0

where *thresh6* is the threshold value estimated from repeated scan data evidence for neuron 6.

This is the operating principle for neuron 6 which we have inferred from the brain-scan evidence. All we can claim is that it satisfactorily "explains" the brain-scan evidence. It constitutes a scientific understanding of how neuron 6 operates as a component of the brain module.

We cannot be sure that it is correct, and we certainly cannot prove that it is correct. Such is the nature of science. It's our best guess that is consistent with all the evidence.

Similar details of operational behaviour can be assumed, and are corroborated by evidence from further experiments, for the remaining 8 neurons in our network. Neurons 7, 8 and 9 appear to behave similarly. Although the "input" neurons 1 through 5 merely pass on the signals they receive, because they each happen to receive only one input, this is the same as each one adding all its input signals together. So all nine neurons can be viewed as behaving in the same way.

Using HYP5 together with further experimental data we can deduce the strength, let's call it the "weight", of each link in from one neuron to another.

Now we are in a position to specify precisely the full information processing principles of our network:

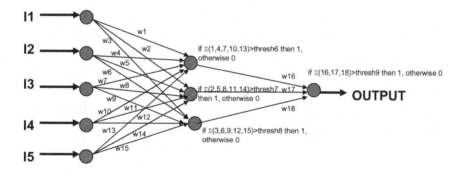

Where *w1* is the value of the strength, the weight, of link 1 and so on up to *w18*, and *thresh6* is the value of the threshold in neuron 6 and so on for each neuron up to *thresh9*.

These principles explain the full pattern of the information processing done by this network. Consider the inputs to neuron 6 whose sum we have characterised as $\Sigma(1,4,7,10,13)$. Now we can be precise about this incoming information along each link. On link 1 the incoming information is: (the output of neuron 1) multiplied by (weight of link 1). This is $I1 \times w1$. Now we can rewrite the processing done by neuron 6 as:

if $\Sigma(I1 \times w1, I2 \times w4, I3 \times w7, I4 \times w10, I5 \times w13) > thresh6$ then 1
otherwise 0

Similarly, this principle of information processing for neuron 6 can be written for neurons 7 and 8. These expressions explain how the internal signals are generated by these neurons; the signals that are their outputs are the three inputs to neuron 9 — let's call them *O1*, *O2* and *O3* for the neurons 6, 7 and 8, respectively.

Now we rewrite the information processing behaviour of neuron 9 as:

if $\Sigma(O1 \times w16, O2 \times w17, O3 \times w18) > thresh9$ then 1
otherwise 0

We know the information processing principles that generate *O1* (this is given above for neuron 6) as well as those for *O2* and *O3* (the similar information processing done by neurons 7 and 8, respectively). So now we can specify precisely, at an information-processing level (i.e., an abstract level above that of neuron pulses and synaptic-gap ion flow), the processing performed by the complete neural network. It is:

if $\Sigma($ {if $\Sigma(I1 \times w1, I2 \times w4, I3 \times w7, I4 \times w10, I5 \times w13) > thresh6$ then 1,
otherwise 0}$\times w16$,
{if $\Sigma(I1 \times w2, I2 \times w5, I3 \times w8, I4 \times w11, I5 \times w14) > thresh7$ then 1,
otherwise 0}$\times w17$,
{if $\Sigma(I1 \times w3, I2 \times w6, I3 \times w9, I4 \times w12, I5 \times w15) > thresh8$ then 1,
otherwise 0}$\times w18$)
$> thresh9$ then 1
otherwise 0

44

For readers with a blank (or otherwise occupied) space in their brains where fascination with mathematics ought to be, this must appear daunting. The advice, however, is (as before) not to look away now, but to read on quickly, trusting that I can do better.

Be reassured, what matters is not the details of the above expression; what matters is the fact that we can plausibly generate such a precise description of the principles of operation of our brain module. All that you need to carry forward is that brain scan evidence may well enable the scientists to figure out the information-processing principles at work within our brains. In addition, you might have some inkling about the sort of inspired guesswork required to develop such principles from brain-scan evidence.

Nevertheless, I'll invest a few paragraphs in an attempt to demystify this unwieldy expression. But those readers who can happily live with this level of mystery are free to skip the next few of paragraphs. Alternatively, relax and simply read on.

In the centre (the three curly bracketed expressions) you see the processing done by each of the three central neurons, neurons 6, 7 and 8, one per curly-bracketed expression. The output signal (which may be nothing) from each of these three neurons is then multiplied by the link weight from that neuron to neuron 9, and all three signals are added up within neuron 9. This is the processing done by neuron 9 and it is described by the initial "Σ" and the three expressions (i.e., $\{...\} \times weight$) within the curved-brackets. This output signal from neuron 9 is then compared with the value of the threshold, *thresh9*, to determine the final output signal from the complete brain module.

Or less mathematically, the above expression for the behaviour of the complete brain module may be summarised as:

if (output signal from neuron 9)
> *thresh9* **then** 1
otherwise 0

or less succinctly and with a small concession to mathematical notation as:

$$\textbf{if} \; (\{\text{output signal from neuron } 6\} \times w16)$$
$$+$$
$$(\{\text{output signal from neuron } 7\} \times w17)$$
$$+$$
$$(\{\text{output signal from neuron } 8\} \times w18)$$
$$> \textit{thresh9} \; \textbf{then} \; 1$$
$$\textbf{otherwise} \; 0$$

I trust that this helps. But, if not, let me repeat that the details don't matter. What matters is that you understand that it is plausible that the scientist may be able to extract from brain-scan evidence the precise details of the "principles of operation underlying"[12] a brain that is thinking.

Next, remember that we have inferred (from scan evidence in conjunction with hypothesised principles of operation) actual values for the 18 link weights (*w1* through *w18* above) as well as for the neuronal thresholds (*thresh6* though *thresh9* above). This leaves the only unknown quantities in this information-processing principle as I1 through I5. But these are precisely the inputs to our network. So when we give specific values to the network's five input signals (I1 through I5), the mathematical version will enable us to simulate the neural network, and calculate its output, which will always be 1 or 0, pulse or no pulse.

What is really important is not what Σ, or any other strange symbol means; **it is that we now know (with reasonable confidence) exactly how this brain module works.**

The operating principle that we have worked out from the brain-scan evidence can be further corroborated by evidence from subsequent scans of the network processing a wide variety of inputs. Extraction and formulation of this information-processing principle has allowed us to accomplish another important step towards a realisation of the Singularity. So, granted a small step into the future to make a few nifty (and light-touch) nanobots available as well as the good fortune to find a small, informationally-encapsulated and non-adaptive module of the brain, the first steps of reverse engineering can become a reality. The "non-adaptive" qualification rules out any learning behaviour because learning would imply that the brain module changes its structure as a consequence of repeated operation, and this would further complicate interpretation of the experimental evidence.

In addition, some knowledge of how neurons tend to behave very usefully constrained our conjectures about patterns of information processing in the data. A quantity of what we might call top-down hunches, if not firmer knowledge, is more than useful to the reverse engineer. In the absence of incredibly lucky guesswork (hardly a viable component of any purported engineering discipline), valid preconceptions about what's going on are essential to reduce the infinite number of possible patterns to a feasibly-explorable few. Otherwise the scientist, even more quickly than the computer, will be stumped by Maxim 10.

For the particular brain within which this module was scanned, we have been able "to see inside, model it, and simulate [one of] its regions... We... understand the principles of operation underlying... thinking, knowledge that will provide us with powerful procedures for developing the software of intelligent machines" (p. 144).

If so, then what's this module thinking? This may be an unfair question, if only because what a brain is thinking may well be a global property of the brain as a whole and not applicable to small bits. After all, you wouldn't ask, "What's neuron 9 thinking?"[13] But because we are aiming to discover the principles underlying thinking we do have to push on further in this general direction and try to make sense of such questions. The route from understanding the information-processing principles of brain structures, and simulating them, to development of the software of intelligence requires a further process of discovery.

Okay, so we have the information-processing principles that underlie our brain module. They can easily be rewritten as a computer program and so be transformed into a software module that can become a component of an AI system — perhaps an intelligent robot. When embedded within a total simulation, can we expect this software module to contribute to the intelligent behaviour of the system? We know it takes in five signals, and we know exactly what it does with them in order to generate its output signal. But here we are stuck. Unless we can trace backwards through the brain's information processing in order to learn something more about where these five input signals originate, and therefore what they represent in terms of the cognitive behaviour of the brain's owner, we're totally stymied. (Alternatively, or better, additionally, we'd like to trace forwards from this brain module's output signal in an effort to learn what it represents at the cognitive level.)

We know what this part of the brain does in terms of signal processing. Each neuron sums its input signals, and outputs a signal if the total is greater than an internal threshold. In addition, we know what these individual threshold values are. But we know nothing about how these signals and operations on them relate to any component of our thinking. We have no idea what the neuronal information-processing principles mean at the cognitive level. Is this brain module dealing with lines, angles, triangles or what? We do know exactly what it's dealing with, and how, but only at the neuro-anatomical level.

Once again, we've crashed into the undeniable truth: **structure does not provide meaning**.

Notice that the science we've pursued so far sits squarely in the domain of the neuroscientist — the expert on brain structures and functioning — so far, there has been nothing for cognitive scientist to get to grips with. Can we get from one domain of expertise to the other? To put it another way: can we move up from the level of brain behaviour, in terms of brain anatomy, to a level of interpretation in terms of cognitive objects and processes, such as knowledge storage and belief management? The undeniable truth asserts that this necessary transition will not be easy, but with some extra help it might be possible.

In order to get a second, and definitely more familiar, insight into this question of the transition between *levels* of understanding a system, we can take a similarly close look at one informationally-encapsulated component of our cognitive lives. Here's the information-processing structure and operations revealed by our (beyond the) state-of-the-art nanobots:

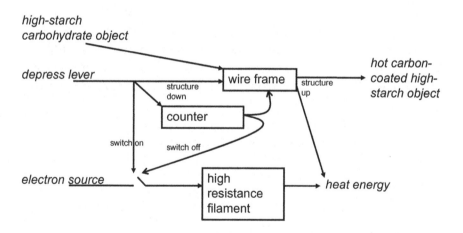

Further than this, the nano-scan reveals that there is a period of electron flow into the device, initiated by depressing an external lever, which continues until an internal counter mechanism terminates it. The electrons mostly go to excite the molecules of a high-resistance filament which then dissipates this excitation as light and heat. The high-starch carbohydrate input is enclosed in a metallic grid directly attached to the external lever. Some of the energy dissipated by the excited filament molecules transforms the surface of the starchy object which is thereby reduced to one rich in carbon atoms. In addition to blocking the electron flow to the filament, the counter component causes the previously depressed lever (and attached metal structures) to rise again.

What is this system? Familiarity with this particular household system may permit you to recognise it. But if I give you the cognitive-level meaning of its three inputs and two outputs, any puzzlement will instantly vanish. Moreover, you will see how to interpret this structural, or syntactic, model at the cognitive level, i.e., in terms of everyday meanings.

The three inputs, from top to bottom, are: a slice of bread, a finger depressing the lever, and a connection to mains electricity. The two outputs are, again top to bottom: a slice of toast and some heat. Those are the useful meanings of the system inputs and outputs. Why are they useful? Because they relate the system to your everyday life in a way that the electron-based atomic-excitation description does not.

The more familiar cognitive level, or semantic, model of the electric toaster is:

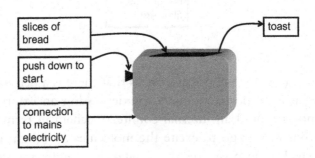

The meanings in terms of electron flow and the action of heat on carbohydrate molecules are just as valid in their own right. But they are not meanings at the normal cognitive level for most people's interaction with this system.

If your goal is to construct a kitchen then nano-scan models of potentially useful kitchen components, such as the toaster one, are not much help. It is the cognitive-level model that is needed because it helps the kitchen builder to relate the toaster's overall function — to produce something edible — and more specifically its inputs and precise output to the overall functions of a useful kitchen. Common kitchen features like slices of bread and a mains power point (not to mention a robot or person to depress the lever) present obvious connections for integrating the toaster system into the complete kitchen.

It is just such cognitive-level meanings that we require in order to get to grips with our brain module. Ideally, we would like to be able to cast the extracted information-processing principles in terms of appropriate cognitive-level objects and actions, i.e., instead of neuronal signals, link weights and summing operations, etc., we need geometric objects, angles and distances, for example.

How can we get from our neuron signal-processing model to its cognitive-level equivalent? Can we cross from a model that is purely structural with respect to the level of understanding we require, to the cognitive-level principles, the content underlying intelligence? Can we build this bridge?

Returning to our toaster system where we have the benefit of both types of model, we can see what needs to be done to bridge the gap between them. Here's a picture of the sorts of relationships that constitute the necessary bridge:

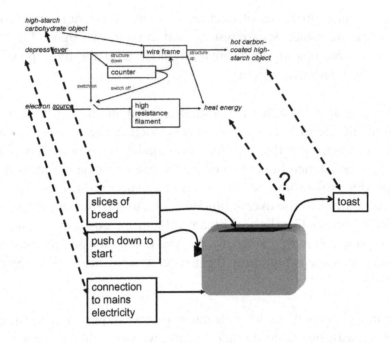

Now we can see the meaning of the structural model in terms of the familiar inputs and outputs of the toaster system. So although structure does not itself provide content, there are ways to make the necessary connections.

One point of note is that the "heat energy" output detected by our scanning bots (i.e., they detected that most of the molecules that constitute the toaster absorbed energy dissipated by the high-resist-ance filament, energy that was subsequently emitted as heat both during, and subsequent to, operation of the electron-stream input) has no equivalent in our semantic model. Why? Because this heat is not a *salient feature* of our toaster system at the cognitive level. Typically, we do not include the hot metal box in our conception of the essentials of a toaster. Ideally, we would choose not to have it. It sometimes makes plucking out the toast a nifty-finger exercise. It is the wasted heat that is generated but does not contribute to brown-ing the toast. Our scanning bots cannot know this. They can only report on the structures and processes they find. Whether what they find is essential or irrelevant to any cognitive-level meaning is inaccessible to the beavering bots.

This is a simple example of the knotty problem of feature salience that is fundamental to the development of an understanding of a system. For the toaster system the heat output is not-salient but it is the semantic model that tells us this. Hence the vicious circularity (Maxim 3) of scanning only the salient features of a brain in order to derive a cognitive-level model that will determine feature saliences.

We have two models of the same toaster system, one in terms of the bare physical components and attributes of the system — the structural model — and one in terms of our everyday interpretation, the "meaning" model. We have been able to construct explicit interrelationships between the inputs and the outputs of these two models. This exercise is fundamental to the development of a complete cognitive-level understanding of the structural model. From these two models it is possible to determine precisely how a toaster works. We can develop this further understanding by relating the internal components and information processing revealed in the structural model to internal components of the semantic one.

With a screwdriver, a pair of pliers and a little DIY competence (after pulling out the plug!), we can reveal the internal components of our toaster. Below we have such an "exploded" model with the further explanatory relationships between our two models made explicit. In the case of toasters we can easily bridge the gap between our everyday understanding — a semantic model, our everyday meanings — and what the scanners reveal about how it is structured and what signals pass internally between the various components, a structural model.

An example of the further, and necessary, understanding of the toaster system that derives from this inter-level bridge building is "heater-element and bread proximity". Without the knowledge that the purpose of the heating element is to brown the bread (i.e., to toast it)

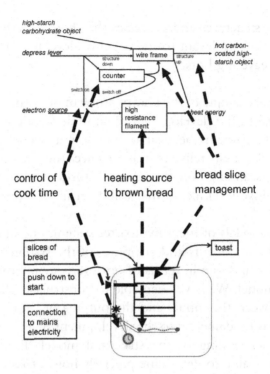

and that proximity makes this process more efficient, then, although the scanners might record the distance between them, they cannot record this "closeness" as a salient feature of the toaster system. It will necessarily be on a par with the many other internal component-separation distances. The "heat generated" feature, which is one of very few outputs, appears to be "structurally" more significant. It is the semantic model that tells us that proximity between heat source and bread is important whilst heat output is not (at least not in terms of toasting although it could be in terms of engineering energy efficiency).

Once more, we see that feature salience and cognitive-level understanding of a system are interlocked. More directly, the salient features *are* the basis of the understanding we seek, and they cannot be determined from a single structural model alone. Hence, Maxim 4: **structure**

does not provide meaning whose truth is reinforced by **the vicious circularity of understanding based on a single example** (Maxim 3).

Successful identification of salient features, the essence of meaning (and hence of understanding), is usually achieved by comparing and contrasting several structurally different examples of the phenomenon being studied. Feature salience cannot be determined from a single structural example. As we shall see, this awkwardness rises in importance when it is intelligence and not toasters that we wish to understand.

Our understanding of the structural model is concentrated in the three central notes, one for each of the relationships explicitly illustrated. In this toaster example, an understanding of the system and hence its role in a kitchen cannot come from the scan-data model alone. This model, which must be restricted to objectively observable phenomena (such as electron flow), does not reveal its operational significance in a kitchen. This is the structure-meaning abyss introduced in the opening chapter.

I should emphasise that my use of the terms "structure", "meaning" and "salient feature", are valid only with respect to the "level" of our enquiry. The so-called structural model of the toaster, for example, might contain significant elements of the "meaning" that a toaster-repair engineer requires. This is because toaster repair is likely to be rooted in electron-flow paths, metallic connectors, springs and the like rather than in slices of bread. At this level of enquiry, the details of the "counter mechanism" can become salient features.

Successful bridge building between our two types of toaster model was crucially reliant on having the two models. In the case of brain scanning for the purpose of elucidating the principles and processes underlying intelligence, we do not have these two types of model. Indeed, the primary goal of reverse engineering is precisely to construct the cognitive-level model. The task facing the reverse engineer is to bridge the gap between two models (as a set of explicit interrelationships) but with little or no knowledge of the cognitive model. It's akin to constructing a bridge from one pier (the structural model) to another of unknown location and structure. That's a bridge-building challenge as we can see.

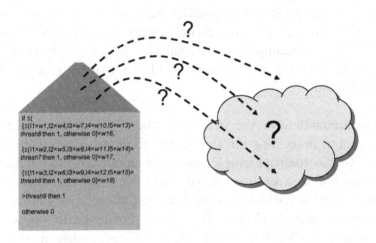

Although the enquiring scientists will, of course, be lacking a semantic model of toaster-like completeness when struggling to understand how brains support intelligence, they may not be as completely clueless as depicted in the above illustration. Indeed, it just so happens that we are not totally ignorant in the case of our little brain module.

Luckily, the brain module we are investigating is not buried within the spatial reasoning area of the brain but sits on the periphery. So it receives its five input signals directly from some external source: five values visually presented, or five values spoken directly, or even five values fed (via the benefits of whatever technology we might need) directly into the neurons 1 through 5.

Now, because we have direct control over the inputs, we will know exactly what they represent, at the cognitive level. We know what they "mean". Therefore we stand a good chance of working out what this brain module is doing with them. Here are the semantic handles that we have on this brain module.

The five input signals are the X and Y coordinates of two points on a graph (two points in a Cartesian space if you prefer a more technical characterisation); call the two points (X1, Y2) and (X2, Y1), together with one further signal, call it L1. So now we've got a foot in the door, as it were, to an understanding, at the cognitive level, of our successful simulation. We know the meaning of four (out of five) of the module's input signals. All we have to do is find the meaning of the module's output signal. What do the outputs 1 and 0 signify in terms of the intelligent activity of this brain, i.e. what do they mean at the cognitive level?

Here's the picture of our new semantic knowledge, the basis for an (incomplete) semantic model:

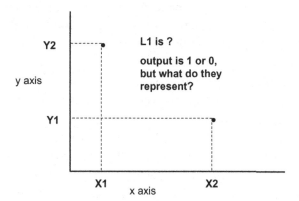

This means that the necessary bridge building is not across to a totally mysterious structure — something somewhere in the clouds. We now have some knowledge about where and to what we are building a bridge.

Here's the more hopeful picture:

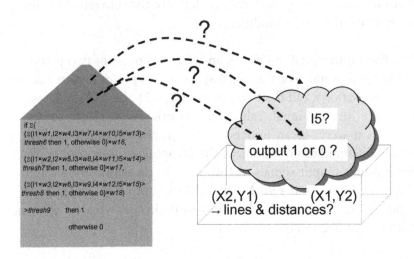

One way to discover the meaning of the module's output is to see how the input signals (most of which we now understand) are transformed. But how are we to make the necessary discovery? We must spot the pattern of information processing behind all the available samples of input signals and paired output signals.

We must search for, and then explore, correlations between the outputs and various relationships between the input values. For example, when all five input signals are large the output is 1; when X1 plus X2 is smaller than Y1 plus Y2 plus L1, the output is 1; and so on. Sadly, there is an infinite number of possible relationships to investigate, and this will take a long time (however much available computer power will have soared up to on its exponential ascent). This means that, in general, such a search is not a practical possibility.

This is another intrusion of our unbelievable truth, Maxim 10: **no computer will ever count 1, 2,... exponentially many**. Exponentially many is, as we shall see, well short of infinity, but quite large enough to defeat all imaginable advances in computer power.

A slice of bread in and a charred one out might well have been enough semantic clues for most of us to realise that the earlier structurally-depicted system was a toaster. This is because, although it is still true that there are an infinitely many possible systems that can accept a slice of bread as one (of three) inputs, we also know the meaning of the crucial output, and there are only a few (maybe just one) normal household systems compatible with the little knowledge made available.

The information-processing model that the AI scientist is grappling with is grounded in the structure and basic operational activities of the original brain module. It must be because the scanning technology can only scan and report on the structures it finds and the ways that they interact. Even light-touch nanobots can do no more than report the structures traversed and the signals that whiz by them.

Brain scanning, it should be evident, can only ever deliver an information-processing model based on the processing units scanned and their interactions — i.e., the structures scanned. How could a brain scanner possibly "see" beyond the structures being scanned? How can the signal-processing devices, our nanobots, get some purchase on what these signals mean? They can't; we must settle for a brain-structure based model, a structural model with respect to our goal of understanding intelligence. This is a model well short of the meanings we need, a semantic model.

Despite the understanding of most of our brain module's inputs afforded by the revelations given above, we can make no further progress in understanding the structural model. In the absence of some sort of more-complete semantic model, such as we have for toasters, we are a bit stuck. We know of no way to bridge this gulf between what scanning and reverse engineering will deliver — purely structural information — and the meanings necessary for building a simulation of human intelligence. This is the abyss between structure and meaning.

Consider the options for configuring a useful kitchen from the purely structural model of the toaster. How could we know that its hot, carbon-coated output should be tied into the process of making toasted sandwiches, and that the heat-energy output is to be ignored (or carefully avoided when fingers are being used to extract a particularly small slice of toast)?

A more plausible application that uses both outputs might be as a component of the kitchen's heating system: the heat energy output will directly serve this role, and maybe the hot carbon-coated object is a canny heat-storage device for delayed or localised heating requirements such as a hand warmer? Without kitchen-level meanings for the inputs and outputs of our device, there is little hope of using it along with other similarly (ill) understood devices to successfully configure a useful kitchen.

The gap between knowledge of the structure and behaviour of brains to that of minds is not going to be bridged by any exponential boom in scanning technologies, nor computer power, nor capacity. Something fundamentally different is needed. But before searching for answers elsewhere we will lay to rest the possibility that developments in scanning technologies, which are moving fast, hold any promise of a breakthrough for the reverse-engineering strategy.

Endnotes

[1] Ray Kurzweil, *The Singularity is Near* (Duckworth, 2005). All subsequent unattributed quotations are from this source.

[2] This is, of course, a crude parody of scientific practice, but nevertheless quite sufficient for the purpose of motivating proper interrogation of AI demonstrations. Apart from the ever-present currents of human egos and the rocks of established science both deflecting and diverting the flow of pure logic, there are unrevealed, wholly rational perturbations. So, for example, a failed experiment may point to an experimental inadequacy rather than a refuted hypothesis, and there may be equally good arguments for both possibilities.

[3] The dealer's rule happens to be only colour based; it is: alternate colour in successive cards. So the underlying rule is simpler than either of the hypotheses. Would the

players ever discover it? Probably, because although the player's focus on the card "features", "face number" and "suit", is misguided, there are not a lot of possible "features" in a card game. The dealer rigged the evidence to be misleading. Nature (we trust) does not do this, but the scientist faced with the puzzle of discovering the principles of operation of the human brain has a much wider selection of potential "features": we will presume that neurons are important but what about features of neuronal information processing? Does the scientist need to hypothesise and experiment in terms of molecular chemistry, or is the higher level of "signal processing" where the useful explanatory rules will be found?

[4] I refer to empirical science (science that tries to develop an understanding of how the world works). In the purely theoretical sciences, such as many branches of mathematics, theories can be proved true, but that leaves the tricky issue of how such theoretical constructs relate to the world you and I inhabit.

[5] Even Sir David Attenborough, although not a scientist but with ready access to the best scientific advice, spoke of Darwin setting out to *prove* his theory of survival of the fittest in *Charles Darwin and the Tree of Life* broadcast by UK 2 BBC Radio 4 at 10 pm Thursday 29th April 2012. Again on BBC Radio 4 *Start the Week*, 9 am Monday 9th April 2012, historian Rebecca Stott talked of Darwin needing to "prove his theory of natural selection". Well, it's still not proven, and never will be. What Darwin did was amass evidence to support his theory, and, more importantly, he found no evidence that refuted it. So often when a scientist is being challenged, we get "but can you prove it?". With the only honest answer being "no", the scientific claim is effectively demolished in the context of the general ignorance of our logical conundrum.

[6] I. J. Good, a colleague of Turing at Bletchley Park, is just one of the prominent scientists to register (long ago) his dissatisfaction with the way science sweeps failed experimental results under the carpet. The highly unsatisfactory practical consequence being that, with no published record of failure, the failed experiments that are based on highly plausible hypotheses may be persistently retried by other scientists. Despite the rock-solid logic, there is never going to be a viable *Journal of Failed Experiments*. Despite a number of start-ups — e.g., *Journal of Errology* and *Journal of All Results* — the negative psychology appears to be triumphing over the sound logic (further complicated by a need for very many speciality-specific versions). The value of successful outcomes rests on repetition across a breadth of topics which is just what AI projects never do.

[7] Although my picture looks crowded, it does in fact illustrate only a dozen or so neurons. Jeff Hawkins, for example, in his book *On Intelligence* (Times Books/ Henry Holt, NY, 2004) written with Sandra Blakeslee states that a 1 millimetre cube of the cortex will contain about 50,000 neurons.

[8] As luck would have it, this module also happens to be a feed-forward network, i.e., no feedback connections between neurons, which is also a requirement for

scan-data interpretation. G. C. Van Orden & K. R. Paap ("Functional neuroimages fail to discover pieces of the mind in parts of the brain", *Philosophy of Science*, 64 (proceedings), pages S85-S94, 1997) stress this necessity (amongst others that we will get to visit in Chapter 3) for the current reality of "subtractive neuroimaging".

[9] "Spurious precision", which means that seemingly different values are in fact indistinguishable, is a common outcome of measurement and calculation. Consequently, the scientist must spend some time on the sensible interpretation of the experimental numbers produced. In our brain-scanning situation, within which both the meaningful values and their modes of measurement are pure conjecture at this time, I ride roughshod over the subtleties and do no more than crudely illustrate this feature of empirical science.

[10] Usually called "rounding", and we will use the particular rounding rule of rounding up: add 1 to the value of the first decimal place if the digits beyond it are greater than or equal to 0.05, and add nothing if not. So, for example, 7.155 becomes rounded to 7.2, and 7.149 becomes rounded to 7.1.

[11] "Thresholding" is nothing more complicated than comparing two numbers (an experimental value and a threshold) and taking the action that generates an output signal of, say, "1" if the experimental number is greater than the threshold, and generates an output signal of, say, "0" if not. More succinctly, we write this process as:

if experimental value > threshold value **then** output 1

otherwise output 0

[12] See chapter opening quotation from Kurzweil's *The Singularity is Near* (Duckworth, 2005), p. 144.

[13] Curiously, the philosopher Bertrand Russell, whilst seemingly challenging the CTM long before its rise to prominence, used football to expose the unfairness of expecting thinking at too low a level. In the1920s, he wrote, "Mental phenomena seem to be bound up with mental structure. If this be so, we cannot suppose that a solitary electron or proton can 'think'; we might as well expect a solitary individual to play a football match." *What I Believe* (Kegan Paul, 1928 edition), p. 12.

ର Chapter 3 ର

Brain Wave Solutions

"There is a circular logic popular in subtractive neuroimaging."
GuyVan Orden and Ken Paap, 1997[5]
"The amygdala... is involved in emotional processing, the hippocampus...
plays an important role in memory, and the striatum... is key in
representing the value of stimuli and actions."
Tali Sharot, 2012[1]

The primary (but by no means the only[2]) sources of evidence for understanding the working brain are the various brain-scanning technologies. So far, we've simply assumed that detailed evidence — primarily the timings and strengths of neuron-network signals — is available and can be taken at face value.

We did, however, reach an impasse when attempting to use this evidence to elicit the information-processing principles underlying intelligence. In subsequent chapters, we will continue to explore possible routes around this block in an effort to attain our primary goal — an understanding of intelligence.

In the current chapter, though, we will take a critical look at this initial component of reverse engineering. We will survey the current realities of brain-scan technologies as a basis for speculating on future possibilities. After all, there is a possibility that improvements in these technologies will give rise to better (and perhaps different) evidence, and that may contribute to a bridging of the structure-content abyss.

These brain-monitoring devices are modern miracles of science but, a little surprisingly, they are also severely limited by systemic weakness when used to provide evidence to support reverse engineering. Once more, this problem is rooted in a lack of perceived joints to carve at (Maxim 1), despite manifest brain joints.

There may be good, if not exactly exponential, growth in the power and precision of the engineering technologies that support brain scanning technologies — functional Magnetic Resonance Imaging (fMRI), Positron Emission Tomography (PET) scans and Event-Related Potential (ERP) techniques.

Tabulated in a 2005 introductory text[3] on the ERP technique, we find the following useful summary of brain-scan technologies (where ERMF is the Event-Related Magnetic Field technique).

	Microelectrode measures	Hemodynamic measures	Electrodynamic measures
Invasiveness	poor	good (PET)	excellent
Spatial resolution	excellent	excellent (fMRI)	undefined/poor (ERPs) undefined/ better (ERMFs)
Temporal resolution	excellent	poor	excellent
Cost	fairly expensive	expensive (PET) expensive (fMRI)	inexpensive (ERPs) expensive (ERMFs)

This tabulation, which goes (left to right) from the precision of single-cell recording through localised collections of neurons to total brain scanning, compactly summarises the current range of choices for brain scanning technologies. Given that scanning nanobots have not emerged, although we are assured that development is well underway, at present we must deal with these three classes, the current realities of brain-scan technology.

As the "invasiveness" row shyly indicates, sticking electrodes through subjects' skulls and into their brains is generally not undertaken without good cause so we can pretty much discount the microelectrode techniques as an everyday technology for investigating normal human cognitive architecture. Much, though, has been learnt using this technology in particular cases of traumatic injury and exceptional mental problems.

In our earlier scientific probing of the small brain module we exploited the "excellent" precision (in both time and place) of the microelectrode measures and yet avoided their "invasiveness". We managed this by assuming the use of mythical nanobots.

Now we will concentrate on the currently available technologies in order to see whether they might conceivably hold some promise of a way around the impasse. At the close of this chapter we'll revisit nanotechnology in order to expose its realities in our current context.

Moving across the tabulation given above, the PET and fMRI scan techniques both suffer from "poor" "temporal resolution" because they measure physiological responses to goings-on in the brain. They both measure increased blood-flow in the brain caused by neuron activity (hence hemodynamic technologies). The increased blood-flow response to neuron activity (i.e., signal processing and transfer) cannot be instantaneous. It takes a little time to build up and a similar time interval to die down again. The neuron-activity patterns are relatively much snappier.

It is like trying to plot the path of a fish by measuring the turbulence it causes in the water. There is both a bow wave and a wake, but from the onset of the first to the disappearance of the second is an (ill-defined) time interval somewhere within which the fish occupied that place in the water. Consider also the potential for further complication when several fish are swimming together but not in a synchronised manner, and the water is also subject to random multiple disturbances throughout.

On the positive side, because the "spatial resolution" is "excellent" we know precisely what brain area is activated, although the best localisations are akin to detecting a massive shoal of fish, never a single swimmer.

Crucially, this technical summary must be assessed in the light of how the various techniques are used. High-precision instruments may deliver equally highly precise measurements, but the *way the instruments are used* determines the limits of the validity of the measurements as evidence.

Instrumented improvements in the precision of water turbulence measures, for example, might permit earlier detection of a fish's bow wave but this will only extend the time interval within which the fish must be placed. It is not obvious that it would improve the precision of actual placement. To do that we might, for example, have to switch to a technology that directly monitors the fish by detecting light reflected from its scales rather than continue perfecting our turbulence detectors.

The basic scan-data evidence is a neuroimage — a three-dimensional picture of the active sub-structures in a brain. With the necessary sensors in place around the head, and the brain stimulated appropriately, a neuroimage can be obtained. Because the generally-usable brain-scan technologies make their measurements from outside the brain, they will always sense neuron activity outside the brain region of interest, and there is always other activity in the brain.

In order to focus the neuroimage on the brain activity derived from the experimental stimulus, a second neuroimage is obtained from the *unstimulated* brain (i.e., of the brain's normal activity in the absence of the specific experimental stimulus). If these two neuroimages are "subtracted" by eliminating all the activity that is common to both images, we are left with an image of just the activity caused by the experimental stimulus.

Subtraction is not a tricky operation: on numerical values (and capitalising on our familiarity with equations), we can subtract 3 from 5 which leaves us with 2, so:

$$5 - 3 = 2$$

Image subtraction is much the same thing. So we might subtract one image from another to get a third, for example:

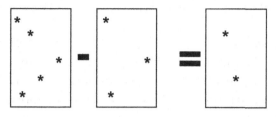

image subtraction

These are two-dimensional images, but subtraction is no different when (as with the neuroimages) the activated areas are dispersed through three dimensions. Again when both images exhibit activation in the same position, that activation gets removed by the subtraction process. In the interest of visual simplicity I'll stick with examples in just two dimensions. So when images are patterns of activation in a brain (designated with "stars"), the subtraction proceeds as illustrated in the next figure:

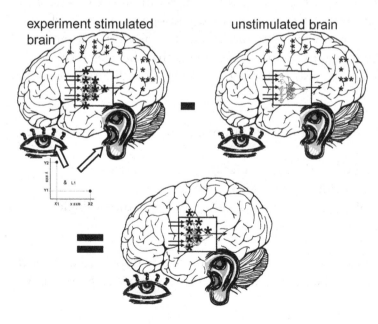

In the final brain-activation image, we obtain a picture of the brain activity that was caused by the experimental stimulus. We see that it is the brain module that we explored earlier (clearly, we're only subtracting the "starred", i.e. activated, elements of the images).

This is quite simple. It would also be valid if the brain was always doing exactly the same thing when scanned with, and without, the experimental stimulus. But, of course, it is not. The reality is more complicated, much more complicated. Brains, both un(experimentally) stimulated and stimulated ones, will always be engaged in many different activities. The result is that the pattern of activity in both sorts of brain neuroimage will never be exactly the same on any two separate occasions, and the activity of interest (caused by the experimental stimulus) must be discovered amid an ever-changing jumble of other brain activity.

Consequently, many repeated "subtractions" are necessary in an attempt to eliminate the activity not due to the experimental stimulus (activity patterns both persistent and transient) and to focus securely on the brain activity of interest[4]. In addition, there is no guarantee that all our brains are structured and operate in precisely the same ways. Consequently, the desire for an objective neuroimage of the selected cognitive behaviour requires that the repetitions and image subtractions also extend to a range of different experimental subjects.

By these means the scientist obtains an "average" neuroimage

- not tied to the idiosyncrasies of any one person;
- free of clutter from other brain activity; and
- similarly free of any one-off peculiarities that a single scan would capture.

This is all good news (apart from the repetitions needed which may be excessive, see next page), but this is still overly simplistic. Why?

There are two further complications:

1. Brain structures are not just active or inactive — there is a range of activity levels apparent in a working brain; and
2. Only the generally-unusable microelectrode techniques can come close to measuring the activity of single neurons — all the other techniques can only measure the general level of activity of groups of neurons.

The hemodynamic techniques are like microphones focussed on a crowd of people to measure the structure of the noise (although microphones have the advantage of measuring noise directly rather than measuring, say, face temperature as a correlate of vociferousness). It is true that neurons, unlike people in a crowd, do not wander about in brains, but crowds are only one layer deep whilst neurons are packed millions of layers deep in brains — not to mention walled in by a thick cranium. This two-dimensional packing difference means that a directional microphone can often be aimed precisely at just one person in a crowd, but brain scanners will always be pointing at thousands of neurons.

Assume that the goal of the crowd scanning is to determine which groups of people make the most noise as a result of, say, stimulating words from a provocative politician. It is true that advances in directional microphone technology now mean that single voices can be picked out in a crowd but not when the crowd is crammed many layers deep inside a thick-walled building and interwoven with a lot of packing. Neural networks inside a cranium and packed around with non-neuron brain cells are just such a crowd.

Recall the reality of brain structure (but still illustrated as a layer only one neuron deep):

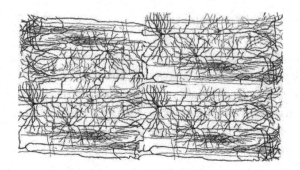

And compare this gross simplification of what is being scanned with the preposterous simplicity of the brain modules that we have attempted to understand:

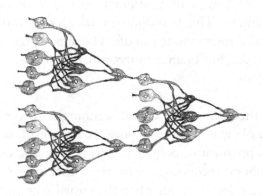

In addition, the first of the two further complications means that the crowd monitoring set-up must set a threshold for the stimulus-responsive "noise". This threshold must separate crowd members who are just chattering to each other (or into their mobile phones) and ignoring the politician's words from those provoked into responses. Any such thresholding strategy presumes that the directly responsive elements of the crowd are the noisiest. In addition, inability to measure the noise of individual people further complicates the outcome because a small, tight group of individual chatterers may produce more noise than a few vociferous responders.

Van Orden and Paap summarised the situation in 1997[5] (and saw no significant amelioration in 2011[6]): There is a "circular logic popular in subtractive neuroimaging... Both require that... modular, cognitive components exist, before the fact, to justify the inference of particular components from images (or other observables) after the fact. Also both require a 'true' componential theory of cognition and laboratory tasks, before the fact, to guarantee reliable choices for subtractive contrasts. None of these possibilities are likely."

In terms of identifying the noisy subgroups in a crowd caused by a provocative political pronouncement, these authors are saying: first, we must assume that such subgroups exist in the crowd; second, we must assume that the chosen announcement does cause the noise elevation that follows it; and third, we must assume that all consistent post-pronouncement noise elevation is a direct reaction to the pronouncement stimulus. Only then can we begin to use subtractive imaging to accurately identify the directly reactive subgroups in the crowd.

The "circular logic" boils down to the observation that the researchers must assume the presence of specifically reactive components of the brain in order to identify brain components that respond to a specific stimulus.

In other words, the brain modules and the modules of our cognitive architecture are both still up in the air, and our best guesses are worryingly interdependent. Neither brain-module activity monitored nor cognitive behaviour singled out is solidly grounded. They rest precariously on each other, disconcertingly reminiscent of a house of cards.

In his 2004 discussion[7] of fMRI and other hemodynamic imaging studies that seek to correlate regions of high levels of brain metabolic activity with particular cognitive functions, Uttal repeats these reservations. In his words, "What is currently missing in the excitement of the new technology is sufficient discussion of the conceptual and technical obstacles that abound in this new endeavour and of possible alternative interpretations of the imaging data." (p. 2)

"Complex interactions across brain regions make it impossible, for example, to determine whether activity in a region is (a) sufficient to support a putative cognitive process, (b) simply a transmission pathway for the flow of critical information, or even (c) an inhibitor of an inhibitor whose surgical removal or activation could produce a simulation of localisation... In short, the foundation idea of cognitive psychology — that the mind is a system of quasi-independent modules that can be localised in narrowly circumscribed regions of the brain — is at least questionable and probably unsustainable." (p. 4)

In other words, Uttal is similarly pessimistic about finding the necessary joints to allow the hemodynamic brain-scan technologies to be focused on "quasi-independent modules that can be localised in narrowly circumscribed regions of the brain".

Moving on to the third column in our table: the ERP technique works differently. It is a refinement of EEG (Electroencephalogram) technology in which the signals from electrodes on the scalp are amplified, and the changes in voltage are plotted over time.

Crudely put, the subject with scalp electrodes attached is stimulated (with say a visual image) and the brain's response to this event is recorded — hence, Event-Related Potential technique. As with the hemodynamic techniques, the raw signals must be subjected to repetition, averaging and filtering (e.g., to remove the inevitable, but essentially random, signal components caused by eye-blinks).

As Luck points out, "we can draw no conclusions without relying on a long chain of assumptions and inferences" (p. 23)[8]. In addition, the

ERP technique requires sometimes hundreds or even thousands of trials to obtain a significant signal, which may or may not be possible, dependent upon task specifics as well as experimental resources.

Finally, the fundamental and persistently recurring issue of tying the recorded signal to (in the case of ERP) a component of a cognitive task remains a stumbling block. "In most cases," Luck writes, "we do not know the specific biophysical events that underlie the production of a given ERP response or the consequences of those events for information processing." (p. 22)

The technologies, though, are improving, and so is our understanding of the biophysics of the brain. Signal processing is getting slicker and timings are being squeezed down so that localisation possibilities are narrowed. But no fundamental shifts, let alone exponential take-offs, in the basic methodologies are occurring, or are expected to occur with these brain-scan technologies.

Also on the positive side there is hope, even "hopeware"; new brain-scan technologies are emerging[9], and so new possibilities for the reverse engineer are sure to open up. But it is hard to envisage how "better" brain scan technology will address the fundamental problem of crossing the abyss, especially when we have little of substance to aim at — due to an apparent absence of far-reaching and fundamental modularity in our cognitive architecture.

The further, and rather different, source of potentially better brain-scan evidence is nanotechnology. Indeed, in the previous chapter we assumed the availability of nanobots in order to give the reverse engineer the very best chance of understanding what the brain module was doing. This is over-the-horizon technology, but how far over is it, given the current time, money and efforts that are being devoted to nanotechnology? We are obliged to dip into this area just in case it will soon relegate the above-described technologies to the dustbin of history.

So what is nanotechnology, and how well is it advancing?

Nanotechnology refers to engineering, such as building machines and robots in its more grandiose conceptions, on the scale of atoms and molecules. A *nanometer* is one billionth of a meter, the width of only three or four atoms. From this viewpoint, everyday engineering is a crude process operating on the scale of billions of atoms, even when dealing with so-called miniaturisation. The possibilities at the atomic level are limitless, but they vary across a spectrum from practical products available today, through research and development of applications reasonably anticipated soon, to blue-sky possibilities that are well over the horizon. Nanotechnological applications from the here and now, and on to the outer reaches of scientific fiction blend seamlessly, so we need to pin down the particular nanotechnology of interest whenever nanotechnology is summoned up as the route to an answer to our problems.

One subfield with immediate promise has capitalised on the progress in genomics. Scientists can now cut, substitute and add atoms to the DNA sequence that is the genome of a life form, and so radically alter the "program" that dictates the organism's properties.

For example, yeast has been genetically engineered so that it produces diesel fuel rather than alcohol. More generally, publicly available "nano-bricks" have been used to construct molecular "engines" that could, for example, roam inside our bodies and can become active to combat specific diseases only when they detect the presence of the disease. NASA claim to be well on the way towards such nano-engineering of astronauts so that their bodies respond automatically to the various unavoidable stresses of living in space[10].

What of the precisely targeted and non-disruptive nanobots that we posited in order to gather the operational evidence from our small brain module? We are most probably looking way over the horizon well into the wide, blue yonder. Why?

At a minimum we must be envisaging mobile nanobots that detect signals and transmit them (as well as a GPS-type location signal, also with nano-precision) to some centralised recording device. Neuronal behaviour of interest — signalling an output pulse or not, and facilitation of signal transfer from one neuron to another — is on the atomic-molecular level. So it is not clear how even a nanobot could be expected to infiltrate a tightly packed mass of neurons, slip between the "insulation" of the surrounding brain cells, and monitor one part of one neuron without disrupting the actions of this neuron. It must also signal exactly which part of which neuron is being monitored — e.g., inputs to the cell nucleus or outputs from it.

But if nanotechnology does manage an unanticipated leap into this particular hard reality over the horizon, then does the reverse engineer have the evidence he or she needs? I do not think so, quite apart from the structure-meaning gap. Here's why.

Reverting to the earlier analogy of trying to detect which members of a packed audience are reacting most forcefully to specific provocative announcements, we must realise that the vast audience is packed millions of layers deep. In addition, each audience member possesses a mobile phone with open conference-call connections to thousands of other audience members. A set-up that manages to monitor the vociferousness of each person individually (as a good correlate of each individual's reaction to the stimulus) will simply swamp the experimenter with evidence. This evidence will vary from valid indications of reaction to the stimulus, to evidence generated by other concerns of some audience members, and also evidence that is generated by any mixture of the two.

This mass of individually equivocal evidence is perhaps why the Singularity doctrine[11] is adamant that the evidence should be of brain regions and not of individual neurons. The averaging effect of brain region activity will go some way towards eliminating the one-off and spurious single-neuron responses. But as we shall subsequently see, the move away from evidence from individual neurons only exacerbates the problems.

In summary, we must concede that brain-scan science (even when broadened to encompass nanotechnology), as currently understood and used, offers no promise of a possible way around or across the gulf between structural principles of brain behaviour and the cognitive-level ones needed for understanding intelligence. A 2013 book[12] confirms the general hyperbole surrounding the claimed abilities of brain-scan technologies and extends the criticisms way beyond our current concerns.

Nevertheless, it would be foolish to conclude that brain-scan technologies (or indeed other techniques as yet undeveloped, or even imagined) will *never* deliver evidence of brain behaviour to whatever precision and accuracy the reverse engineer desires. Human ingenuity is not circumscribable. It would, however, be equally foolish to declare that advances in our current technologies (even exponential advances) and the ways they are used are going to solve anything fundamental with regard to the reverse engineer's task.

The essence of this impasse confronting the neuroscientist seeking an understanding of intelligence is captured in a lack of joints for science to carve at (Maxim 1), coupled with the absence of a bridge from structure to meaning (Maxim 4).

Settling for improved structural models of brain behaviour is the best we can hope for from improvements in brain-scan technologies. However, perhaps the abyss can still be bridged if we can develop some constraints on our target models on the far side — i.e., some principles and guidelines for the cognitive models we seek. If we can discover anything about the structure and operation of the possible modules of cognition, then we will have something to aim at. This must be more promising than hypothesising into the unknown with no clear ideas about the models sought. It is true that Christopher Columbus succeeded in discovering a new world by launching into the largely empty sea but he did know exactly what he was looking for — terra firma. What's terra firma for the reverse engineer?

Endnotes

[1] Taken from a summary section towards the end of the cognitive neuroscientist Tali Sharot's book *The Optimism Bias* (Robinson, 2012) p. 205.

[2] As Victor Johnson, ERP expert, points out (personal communication 28th February 2012), "There is much more to be discussed [in addition to brain-scan technologies],...

1. We are not the only intelligent beings on the planet; so examining learning, memory, concept formation, etc., in other animals provides insights into the fundamental processes required for human intelligence... Almost everything we know about the organs of our body (livers, kidneys, heart, etc.) came from animal studies; the brain is no different. Also, animal studies permit the use of techniques that would be considered unethical for human research (knock-out genes; cortical and sub-cortical lesions, chemical blocking of neurotransmitter pathways, etc.)...

2. Human accidents, neural disorders, and chemical interventions are other sources of knowledge. For example, accidental frontal lobe lesions... help to clarify the functions of these neural regions, and understanding the psychopharmacology of different general anaesthetics (drugs that dissolve consciousness) is probably the most important clue for defining the neural mechanisms responsible for consciousness."

[3] Steven J. Luck's book *An Introduction to the Event-related Potential Technique* (MIT Press, 2005).

[4] A further difficulty caused by repetition of a stimulus is that the brain is notorious for "habituating" to repeated non-significant stimuli. In plain language this means that we humans soon block out any repeated stimulus that is not important to us. Every reader is likely to have personal knowledge of how they no longer (or only occasionally) hear a clock that chimes every hour. This is because the brain ceases to respond as brain scans show — it habituates.

[5] There is a "circular logic popular in subtractive neuroimaging and linear reductive psychology [which assumes that simple elements of behaviour each correspond to a cognitive component]. Both require that strictly feed-forward [i.e., no loops back in the module], modular, cognitive components exist, before the fact, to justify the inference of particular components from images (or other observables) after the fact. Also both require a 'true' componential theory of cognition and laboratory tasks, before the fact, to guarantee reliable choices for subtractive contrasts. None of these

possibilities are likely. Consequently, linear reductive analysis has failed to yield general, reliable, componential accounts."

So wrote Guy C. Van Orden and Kenneth R. Paap in "Functional Neuroimages Fail to Discover Pieces of Mind in the Parts of the Brain" in *Philosophy of Science, 64 (proceedings)*, pages S85–S94, 1997.

[6] Personal communication from Ken Paap, January 2011.

[7] In an article entitled "Hypothetical High-Level Cognitive Functions Cannot Be Localized in the Brain: Another Argument for a Revitalized Behaviorism" published in 2004 in the *The Behavior Analyst 27*, pages 1–6 No. 1 (Spring) by William R. Uttal.

[8] Steven J. Luck's book (Endnote 3).

[9] Worthy of consideration are the so-called "causal neural methods": transcranial Direct Current Stimulation (tDCS) is a prime example. It is a non-invasive method that uses a rapidly changing magnetic field to cause activity in specific or general parts of the brain with minimal discomfort. This facilitates study of the functioning of components of the brain, and interconnections therein.

[10] These examples were featured in the UK television BBC 2 *Horizon* programme *Playing God*, broadcast at 9:30 pm on Tuesday 17th January 2012.

[11] Ray Kurzweil in *The Singularity is Near* (Duckworth, 2005) repeatedly stresses the necessity to build models above the level of individual neurons. For example, "In order to reverse engineer the brain, we only need to scan the connections in a region sufficiently to understand their basic pattern. We do not need to capture every single connection." (p. 166)

[12] *Brainwashed: The Seductive Appeal of Mindless Neuroscience* by Sally Satel and Scott O. Lilienfeld (Basic Books, 2013) details the technology of brain scanning (or "brain scamming" as they explain some of the marketing) and precisely why, and in what ways, it is misused to explore what we are thinking.

∽ Chapter 4 ∾

Whole Parts of Minds

"Given our staggering ignorance of how, in principle, the cognitive mind might be organized, this observation [that our cognitive architecture cannot be entirely modular] doesn't help a lot to narrow the field. But maybe it helps a little."
Jerry Fodor, 2000[1]

"Any effort at neurological or cognitive reduction into separable brain modules or cognitive components may be a search for a chimera."
William Uttal, 2004[2]

"The limits of modularity are also likely to be the limits of what we are going to be able to understand about the mind, given anything like the theoretical apparatus currently available."
Jerry Fodor, 1983[3]

"The condition for successful science… is that nature should have joints to carve it at: relatively simple subsystems which can be artificially isolated and which behave, in isolation, in something like the way that they behave *in situ.*" So said philosopher Jerry Fodor in *The Modularity of Mind* (p. 128)[3].

In this chapter we'll tackle this "joints problem" head on: how and where might the scientists — neuroscientists and cognitive scientists — find modules to study? Where are the joints to carve at in the brain and in the mind?

Why this concern with finding modules? Because "divide and conquer" is as vital for the scientist trying to develop an understanding of intelligence as it was for Julius Caesar trying to convince the truculent tribes of Germany and Gaul of the benefits of Roman civilisation.

Fodor encapsulated his musings in *The Modularity of Mind*, a 1983 book[3]. As you can guess (if you've picked up on my overall sympathy with many of his views), his conclusions are none too positive.

He explores the various ways that minds might be modular — two major possibilities are "vertical modularity" (modules focussed on specific topics, such as vision or language), and "horizontal modularity" (modules focussed on, say, memory or belief management, cutting across specific topics). His analysis, which is generally not supportive of informationally-encapsulated modules in the mind, does hold out the possibility of vertical modules at the point where information first enters our brains — he calls them "input modules", and they might account for the first stages of processing in both vision and language as well as in the other sensory modalities.

Crudely summarised below is my interpretation of Fodor's most positive view of the modularity of minds:

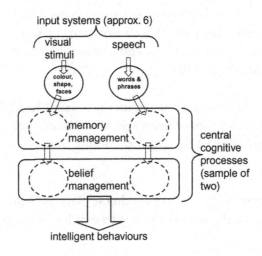

Beyond these, though, he finds no evidence for modularity and this, he notes, is likely to be the limit of our understanding of any natural phenomenon, because the scientist "needs joints to carve at." Hence he concludes:

> *"The limits of modularity are also likely to be the limits of what we are going to be able to understand about the mind, given anything like the theoretical apparatus currently available" (p. 126).*

Via the possible promise wrapped up in the final phrase quoted above, I shall make room for a marginally more positive stance. But, for now, we'll continue to probe for possible joints.

Even the partially modular mind scheme illustrated above looks to be based more on hopeful optimism than empirical evidence of human brains at work. Sadly, something closer to a reality may be what Fodor calls the "global" as opposed to the "modular" mind. Here's my version of his global mind:

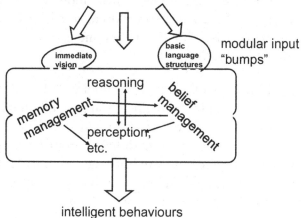

As you can see, not much modularity is apparent. There are no useful chunks for science to work on in its classical manner. This belief is enshrined in our Maxim 1.

Some vestigial modularity centred on the first stages of input processing is about all that might be prised loose and studied away from all the complexities of the rest of cognition. Decades of failure with respect to the Turing Test projects, for example, supports this view. A move onwards from word recognition (which textual input dodges) to competence in human language quickly gets sucked into the central morass of reasoning, beliefs, perception, etc.

An additional problem, an extra worry that Caesar was spared, is that of the relationship between any modules carved out of the two primary levels of modelling — brains and minds. Ideally, the modules should match one to one: identifiable regions of the brain account for components of intelligence — components such as learning, or reasoning, or calculation, or geometry, or whatever. In the worst case, the ostensible brain modules fail to exhibit any significant one-to-one inter-level relationships with hypothesised modules of cognition.

It's as if the individual tribes were not the basic modules that Caesar needed to manage (i.e., co-opt or conquer) in order to enjoy the benefits of Roman rule. Suppose the crucial modules turned out to be tribal shamans, or fathers of two children, or any other grouping that both cuts across the different tribes and is non-obvious to a would-be militant benefactor?

For example, the process of learning, an undisputed fundamental of intelligence, may manifest in the brain as a suite of differently specialised mechanisms, each distributed throughout a variety of brain regions. Learning language may be different from learning mathematics which in turn is different from learning new faces, and all three may each be distributed within different but partially overlapping brain regions. In this case, a cognitive module that captures the common essence of human learning behaviours (if there is one) cannot be attributed to the behaviour of any one region of the brain.

The spectre of a lack of correspondence between the components of brains and minds (assuming we can find some carveable joints in both) is real. How come?

Surely, we take the brain to be the product of long-term local adaptation as evolution theory requires. It is true that we know nothing of the cognitive architecture of the ancestral ape, but evolutionary timescales do seem to far outstrip those of cognitive development, as far as we know. It does appear that our cognitive abilities developed and blossomed far more quickly (tens of thousands of years) than the time span of radical evolutionary change of brain anatomy (millions of years).

As Fodor observes, "Our brains are, at least by any gross measure, very similar to those of apes; but our minds are, at least by any gross measure, very different. So it looks as though relatively small changes to the neurology must have produced very large discontinuities ("saltations", as one says) in cognitive capacities."[4] Based on current knowledge, such an argument can be only speculation.

I can think of no compelling reasons to expect that evolution would deliver a system that in any way echoes rational thought processes, even though the system in question is itself responsible for the rational thought processes. Is a good reason to be positive the fact that many of evolution's other creations, life forms in general, do appear to be readily amenable to science's preferred mode of operation — i.e., carving off components that have limited interactions with the remainder of the system?

In *The Blind Watchmaker*, Richard Dawkins takes pains to explain why step-by-step evolution can be expected to deliver complex life forms that do have joints for science to carve at.[5] For Dawkins, the main point is that such a structuring of complexity is not evidence for the influence of a designer. It is what we should expect from evolutionary development.

For us, Dawkins' argument, which downplays saltation theories, might be taken as support for a similar decomposability of the intelligent mind. This would be encouraging for the cognitive scientist searching for manageable chunks to carve out, except that Dawkins concentrates on the structure of physical complexity. In our case, this is the brain, which does exhibit distinct components (albeit tightly entangled). Our concern, however, is the architecture of the mind. Does the decomposability of the physical translate into a similar decomposability of the mental? Have we any justification for assuming it might?

All this is rampant speculation. It does, however, serve to remind us that the structure of the human brain and its cognitive architecture may exhibit no straightforward mapping from one to the other. If this were the case to any significant degree, then the reverse-engineering strategy takes another big hit.

But first, we need to find some joints to carve at. After that we can worry about what, if any, inter-level relationships they might support.

Here's a positive start: the optic tract is a tangle of neurons that connects our eyes to our brain proper. It has long been known to deal with some of the basic image processing necessary to transform a raw retinal image — rather like the array of pixels that is a digital camera image — into the elemental components of what we perceive. At a minimum, it appears to deliver lines and primitive shapes to the brain for final interpretation as the objects and scenes that we perceive when we look around at our world.

Optical illusions provide a wealth of examples that all of us can appreciate. They constitute evidence of some modular partitioning within our minds, and hence perhaps also our brains. Consider this image:

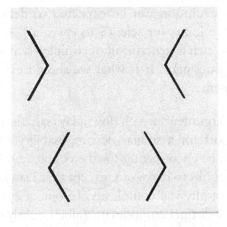

You can see that there is clearly a bigger gap between the points of the top (inwardly pointing) angles than between those of the lower (outwardly pointing) angles, right? Wrong. Let's add two parallel lines.

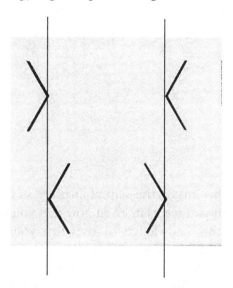

Ah! So the gaps are the same. Are they? Well, if the two vertical lines are indeed parallel, the gaps must be the same, but they still don't look to be exactly the same. Maybe the lines are not exactly parallel? Maybe they slope slightly inwards towards the bottom end? Your mind continues to insist that these gaps are not quite the same size, but reasoning (based on the parallel lines) tells you that they must be — unless the lines are not quite parallel.

It is as if what you see cannot be overruled by what you know. Your visual perception module appears to be isolated from your knowledge of the "reality" and the belief it engenders. You see the gaps as different but you know they are identical. Your visual perception module appears to be *informationally encapsulated*, certainly with respect to your reasoning and the beliefs it generates. This means that your reasoning about these images cannot penetrate, and so influence, your visual perception module that is in charge of telling you what you are seeing. Let's try one more example.

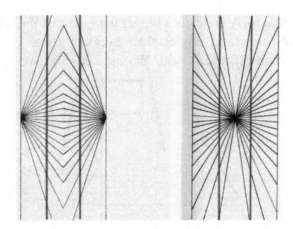

In both parts of this image, the central lines are straight and parallel, but that's not what you see, is it? Even now that you know the reality, you cannot get this knowledge to overturn your visual system's interpretation of what you are seeing[6].

What you see in these images, despite knowing that it is inaccurate (and why it is inaccurate), is suggestive of a brain module for basic image interpretation that is informationally encapsulated, i.e., the brain region where your knowledge of geometry (at least) is managed has no connections into it. Perhaps we've found a joint! One that would permit the scientists to carve out and study an image-interpretation module as a component of the mind?

On the other hand (and there always is one other hand, at least, to be considered in the workings of the brain), occasionally when heading towards Exeter on the A38 in Devon, England, I have crested Telegraph Hill to see the Exe Valley laid out below me and full of water. This sight is always surprising because the River Exe is not normally visible. It is typically the city of Exeter spread out in the valley below that meets the eye at this point in the journey.

Has the water level risen some few hundred feet, and so flooded the valley? That appears to be the situation but my knowledge of clouds hanging low in valleys, my beliefs about the improbability of such a massive rise in water levels unreported on the news, and so on, soon force a reappraisal of what I am seeing. Quite quickly my perception of the scene below me switches to one of a cloud-flooded valley. This is still an unusual interpretation of the scene, and an odd sight, but one that my knowledge tells me is more likely than my first impression, and I quickly see it as the reality before me.

Everyone is familiar with such initial misperceptions, often in half-light or otherwise poorly or unusually lit situations. An essential element of these erroneous perceptual events would appear to be limited information, i.e., your perceptual processing mechanisms have to come to a decision about what you are seeing on the basis of poor quality input to your eyes — edges, shapes, etc., are not well-defined in the half-light.

What you first see turns out to be wrong. Sometimes the reassessment happens fairly quickly and appears to be triggered by knowledge of how the world must be, as opposed to what you seem to be seeing. In short, your visual perception module (if you have one) is sometimes seemingly quite open to influence from your knowledge and beliefs. Does this mean that it is not informationally encapsulated at all?

Further than this, the common, usually transitory "train illusion" can occur on a bright, sunlit day. You are sitting on a train in a station, and suddenly you see an adjacent train and sometimes even the adjacent platform both begin to move slowly past you. Of course, it is your train that is beginning to pull out of the station, but in the absence of any feeling of motion, that is not always what you first "see". But if your train starts with a jolt, this visual illusion is unlikely to occur which suggests that the relevant perception module (or modules) also admit information from other sensory modalities (the feeling of a jolt).

But given that there are compelling, and persistent, visual illusions, both you and I have some reason to believe that our minds (and perhaps our brains) are to some degree partitioned into modules that sometimes operate more or less independently of our "reasoning abilities" at least.

It is an issue that has long been explored and debated way beyond the concerns of building simulated brains in the next few decades. Much less ambitious aspirations, usually focussed on developing theoretical aspects of human cognition, have wrestled with how the cognitive mind could be structured so that it exhibits the amazing behaviours we call, collectively, intelligence.

Singularity enthusiasts really are committed "brain men", and why not? This organ is the generally agreed basis for human intelligence; it is a tangible object directly amenable to scanning, and one day it may well be an open laboratory for nanobot roaming. One of their big mistakes, however, is to jump from the agreed truth that brains cause minds — i.e., our cognitive abilities are provided by our brains — to the unwarranted conclusion that scanning brains at appropriate levels will deliver most of the knowledge necessary to build a mind. Curiously, they never say anything like "build a mind". In 652 pages, Kurzweil never uses the term "cognitive mind", and "mind" only seems to intrude on a couple of occasions via other people's quotations. Why is this?

"Mind" can be a slippery concept as philosophers have amply demonstrated over the years. But we only need to employ a limited and relatively straightforward use of this word. We are not concerned with, for example, consciousness — whether our brain model will be conscious (as it might claim it is). Such things would be nice to know and interesting to debate, but let's get on with building models to facilitate understanding of our basic cleverness first.

For the moment (and the foreseeable future), "mind" or alternatively "cognitive mind" are just labels for the collection of behaviours, such as reasoning and pattern recognition, that we call "basic intelligence". To put it in this way: it is the structure of the mind, the architecture of cognition, that we want to discover, and that the Singularity seekers need as a first step towards their grand goal.

Okay, so we're using the term "cognitive mind", and hence "cognition", unencumbered by all philosophical baggage that gives endless employment to philosophers. We'll assume that this can be ignored (to a first approximation, at least) by those of us who *merely* want to build a basic model of human intelligence in terms of whatever substrates prove to be most convenient and powerful.

So what do we know, or think we know, about the structure and sub-functions — the architecture — of human cognition? If reverse engineering is ever going to get properly off the ground and racing up the exponential curve, then such guidance grounded across the abyss will be essential for the interpretation of the scan data. So the more we know about the possible architectures of the target information-processing models the better.

Singularity seekers accept the need for such guidance and hence overplay (as they must with the demanding timetable they've set themselves) the current state of traditional AI — the bulk of the available models of various aspects of cognitive behaviour. These will be their sources of guidance and constraint on, and augmentation of, their brain-based models. We'll pick up this dodgy thread in a later chapter. In this chapter we'll stick with the long view which is undoubtedly more nebulous but free of fixation on any specific set of models.

Leaving the brain aside, for the moment we'll focus on the architecture of cognition, both its modularity and the internal structure of the modules. How and in what ways is our cognitive architecture, our mind, modular?

It's all very well, and superficially persuasive, to talk of discovering "the principles of operation of human intelligence" (p. 293), and of "understanding their [the biological neurons'] key information-processing methods" (p. 444), and it is especially comforting to read that this is "well underway" in the time-honoured AI tradition. But in the absence of any details about these principles and key methods — how they might be structured, and how they might function — the poor reverse engineer is left floundering in a vacuum of ignorance where the necessary assistance ought to be.

Can we help? What do we know about the modules of cognition? Nothing much is widely agreed. The relevant experts can, and do, hold widely different opinions. How can this be? The cognitive mind, and thus its architecture, is, of course, an intangible system. We can probe the causal organ directly through brain-scan technologies, and perhaps the mind itself more or less directly through psychological experiments. Subsequent analysis of the evidence and its various implications leads to incomplete and contradictory theories.

No detailed theory about the modularity of the mind bears up well under close scrutiny. Hence, Fodor's chapter-opening quotation asserting "our staggering ignorance" about modularity issues which does not bode at all well for the reverse engineer. However, absence of a complete and widely-accepted theory does not necessarily mean there are no respectable hypotheses about what goes on inside the cognitive mind. There are hypotheses about the structure of thought. And any clues must surely be better than nothing at all.

First, the modularity issue: perhaps we just started too small with the nine-neuron brain module? What general "principles of operation of human intelligence" could we expect to extract from a tiny module? One that delivers only a small, and quite specific, component of our spatial reasoning competence?

A brain composed of such modules would need hundreds, perhaps thousands, of them just to account for a competence in basic geometry. If our little brain module was indeed entirely informationally encapsulated (an assumption that greatly facilitated the interpretation of brain-scan evidence), then it had to *contain within itself* whatever general knowledge it employed (e.g., knowledge of X-Y coordinate systems). Extend this supposition (one pretty much demanded by reverse engineering) to a whole brain and what do we end up with?

Anyone with a geometry-competent brain can use the basics (at least) of Cartesian space — i.e., points within X-Y axes defined by an X-value and a Y-value. The owner of this brain can also apply this basic knowledge in countless specific situations, and augment it as directed by experience.

But how can this unchallenged (I assume) cognitive behaviour be explained if our subject's knowledge of geometry is dispersed in the very many specific interpretations each totally cut off (because each is informationally encapsulated) from every other one?

There is a fundamental conundrum here. How can this brain be said to have a general knowledge of Cartesian space when it manifests as separate bits and pieces resident in modules that (by supposition) do not inter-communicate? Our encapsulated modules communicate only via their specific inputs and outputs. If these explicit channels also handled learning and global consistency of knowledge, which might be just about plausible, then the complexity of the module's behaviour takes another big leap forward, one that destroys information encapsulation.

So, the absence of inter-communication we refer to here is that of our brain module's "knowledge" of Cartesian geometry (some of which we presume it must have in order to process the X-Y-coordinate input signals). This particular knowledge, whatever it is exactly, is contained within the module and is not communicable to or from any other brain modules. This is what informationally encapsulated means, and it is this constraint that offers hope for the interpretation of brain-scan evidence by the reverse engineer.

A more plausible (but not necessarily more real) alternative is that knowledge of Cartesian geometry is a relatively global property of this brain. Is it general knowledge that can be applied in many different specific circumstances of geometric reasoning, such as in the brain module we studied earlier?

This explanation sits more comfortably with observations of human geometry competence — for example, if your knowledge of X-Y coordinates is modified, this modification is instantly operative in a wide range of modules that use this general knowledge. This observation is at odds with the encapsulated-module standpoint.

Here's a picture of three (variously unsatisfactory) possibilities:

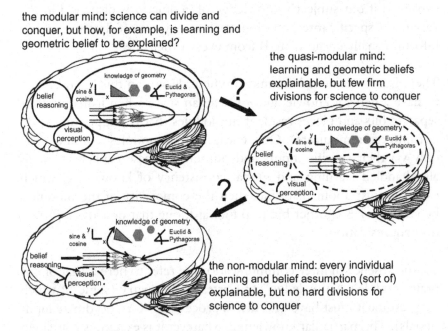

the modular mind: science can divide and conquer, but how, for example, is learning and geometric belief to be explained?

the quasi-modular mind: learning and geometric beliefs explainable, but few firm divisions for science to conquer

the non-modular mind: every individual learning and belief assumption (sort of) explainable, but no hard divisions for science to conquer

This impasse between informationally-encapsulated modules (the essential desideratum for reverse engineering and highly desirable for any cognitive science), and globally accessible general knowledge can, of course, be fudged to some degree. These two extreme mind-plans are illustrated within the top and bottom brains, and a fudge is in the middle.

As every reader can appreciate, a non-modular mind model sits most comfortably with their apparent ability to think quite seamlessly between topics and beliefs about those topics. If, for example, I point out that a slice of cheese is often triangular, you will have no trouble understanding what I mean (even if you disagree with my observation).

Yet if we're betting on mind modules and the consequent restrictions on information flow between them (as science seems to need), then a close coupling between your knowledge of geometry (which is presumably where your knowledge of triangles resides) and your knowledge of cheese is unlikely to be one of the first exceptions you would concede.

Maybe this is because your knowledge of cheese already includes knowledge of the triangularity of cheese wedges. Suppose this is correct, and further suppose that we continue to discuss this fascinating observation about cheese shapes.

You might point out that although you can see what I mean, cheese more often than not is cut from a wheel. So the wedge is not a triangle because one side is curved. It is, you tell me, a sector of a circle. I concede your point and learn this new fact about cheese wedges. I also learn about the name of a "triangle" with one side curved — it is called a "sector".

But where in my brain does this new knowledge go? It must go into the cheese-knowledge module otherwise I'll take you through this same line of reasoning every time the cheese-wedge issue arises. However, it would also appear to be general geometry knowledge as well, so that module must be similarly updated. And so must every other module that might reasonably touch on triangles and circles, such as all my fruit-pie modules.

This is ridiculous for at least two reasons: there will be an endless list of specialised knowledge modules that will require updating, and if these modules are informationally-encapsulated, there is no way to insert the new information anyway.

Indeed it is difficult, if not impossible, for you to come up with two unrelated topics. Try it. The connections just spring out as soon as you bring the topics into focus in your mind. The extreme of informationally-encapsulated modularity is, on the face it, not a promising possibility, but the scientist must follow the hypotheses objectively and not be swayed by introspection.

There are many intermediate standpoints (the middle brain in the earlier picture illustrates just one) between global to the entire system and local to just one module. But once we relinquish total encapsulation of brain modules, we do more than put another dent in the reverse-engineering enterprise. We hole it below the waterline. This is because analysis of scan data from an *un*encapsulated brain module will be further complicated by signals coming in from (and going out to) other regions of the brain.

The divide-and-conquer strategy for understanding intelligence, as well as for subduing ungrateful tribes, has been more graphically stated elsewhere. To fill out the quotation that opened this chapter, "The condition for successful science... is that nature should have joints to carve it at: relatively simple subsystems which can be artificially isolated and which behave, in isolation, in something like the way that they behave *in situ*. Modules satisfy this condition;... wholistic systems by definition do not."[7] Fodor's analysis leads him to suppose that central cognitive processes (such as our belief system) are non-modular which, if true, is (as he says) "very bad news for cognitive science".

It is, perhaps, even worse news for those hoping to reverse-engineer their way through the Singularity plan, because that project is predicated upon achieving a *total* simulation. Anything less scuppers the grand goal of "humans transcending biology"[8]. The freelance cognitive scientist, however, can model whatever turns out to be reasonably modular, and just chip away at the central tangle. It is not a similarly all-or-nothing enterprise (it is also not expecting to solve the problem of understanding cognition in the next couple of decades).

Yet no one, I guess, expects these modules to be anything like totally informationally-encapsulated. In the previous chapter, two independent summaries, both highly critical of the use of brain-imaging data (using functional magnetic resonance imaging, fMRI, and positron emission tomography, PET imaging, respectively), conclude pessimistically that the necessary cognitive modules do not appear to exist[9]. One, by Uttal[10] in 2004, goes so far as to say, "In short, the foundation idea of cognitive psychology — that the mind is a system of quasi-independent modules that can be localised in narrowly circumscribed regions of the brain — is at least questionable and probably unsustainable."

On a brighter note, analyses and explorations of cognitive behaviour do tend to support the notion of modular peripheral cognitive processes, at least. The earlier optical illusions may just be the outcome of such modularity.

In support of the reverse-engineering strategy, it might then be these peripheral processes of cognition that are most easily mapped to brain structures — regions responsible for visual competence, for auditory processing, etc. As another example, you might consider your ability to "hear" individual words in a continuous stream of your native language and contrast this with the extreme difficulty of such perceptual partitioning when listening to utterances in a foreign language.

This lucky break with what we might term, initial perception, is quite consistent with our earlier reservations about isomorphisms between brains and minds. It is exactly our peripheral cognitive behaviours that are found to be as well developed (if not better developed, e.g., the sight of vultures) elsewhere in the tree of life; they are not the relatively recent acquisition exclusive to humans whereas sophisticated learning, knowledge management and belief systems are (or so we fondly believe).

So perhaps there's some hope for reverse-engineering some cognitive modules. But, if so, do we know anything about the finer structure of cognition — the internal structure of these modules, the "structures" of our thinking, the structures underlying the principles we seek? Do we have any credible hypotheses about our targets across the abyss? Maybe we do.

It has long been argued by Fodor, among others, although by no means universally accepted, that human thought processes must by structured like human language. After all, if a major function of our language is to express our thoughts, which is arguably true, then it is not unreasonable to suppose that the structure of one mirrors that of the other. Our language structures are objectively observable, so perhaps they give us an insight into the structures that underlie our thinking?

For example (to run roughshod over one of Fodor's illustrations), if I assure you that *John and Mary like cheese*, and then I ask you, *Does John like cheese?*

Quick as a flash you will answer, "Yes, of course he does, you've just told me that."

But I didn't tell you *John likes cheese*; I told you *John and Mary like cheese*. These are not the same statements, and I've only assured you of the second one. So why do you think that the first one is also true?

The reason why you profess a knowledge of John's fondness for cheese is, so the argument goes, because your beliefs are not indivisible lumps. They have an internal structure — your overall belief about John and Mary is *composed* from more elemental beliefs. Your derived belief that *John likes cheese* is a *component* of your initial belief that *John and Mary like cheese*.

96

To explain further: the suggestion is that the first belief is *composed* from beliefs about two individuals as well as beliefs about other components, each of which is most conveniently tied to one of the words in my statement of the belief. And moreover, for human cognition to work effectively (as it does) over a vast range of these types of belief, the underlying mechanism really ought to be *systematic* just as human language is — e.g., a belief in (**A** and **B**) normally requires the same belief in both **A** and **B** separately, in all similar contexts whatever the beliefs **A** and **B** happen to be.

Suppose I tell you I believe that global warming is a serious threat and that politicians will not tackle the problem. You might then ask me if I really believe that global warming is a serious threat. I might then reply, no, I don't believe that global warming is a serious threat. A suspicious look and even a little backing off to put some distance between me and you would not be an unreasonable response to my peculiar behaviour.

The conjunction "and" is part of an accepted system for conveying complex thoughts (pretty much) regardless of the component thoughts so conjoined. Use of "and" as well as other joining words and punctuation tend to have a fixed meaning, and much the same one among same-language users. We are presuming that an analogous similarity holds among the thought structures of that group. This is not to say that confusions and misinterpretations of articulated thoughts do not occur. It is merely to say that mostly, we operate under a relatively stable and very similar set of rules. The undoubted benefits of such a common, systematic approach to language are expected to occur, and be similarly exploited, in the structuring of our thoughts[11].

Systematic compositionality also accounts for our *productivity* — our ability to generate seemingly endless belief structures (and language utterances). This mapping between the structure of reasoning and the structure of language carries the further implication that such reasoning is primarily *local*, i.e., it is the immediate constituent elements of a thought (the analogues of the words) that largely determines the belief we attach to the thought.

How could it all be otherwise? Quite easily, as we shall see when we explore holistic representations of knowledge. For present purposes, we note that *local systematic compositionality* has a good claim to be a fundamental feature of the thought processes that give rise to intelligent reasoning. This is the bare bones, if that, of the Language Of Thought (LOT) hypothesis, argued at length in a book of that name by Fodor in 1975[12].

To reiterate: if one of the primary purposes of language is to express and so communicate our thoughts, then it is not all surprising that the basic structure of our language productions — both spoken and written — mirrors that of our cognitive processes. Plausibility, though, is far short of a clincher in science, but given no viable alternative, we'll push on with Fodor's LOT hypothesis. I use LOT, however, as merely convenient shorthand for the general hypothesis that the architecture of cognition exhibits *local systematic compositionality* whatever more detailed forms the various specialists might favour.

Your complex thoughts are structured rather like your complex sentences, i.e., complex thoughts are composed of more elemental beliefs that are put together using generally accepted rules. So any belief, or thought, involving *John and Mary* is composed of the same belief, or thought, about both John and Mary separately. It is the "rule" about the meaning of the component "and" that dictates this implication.

In a nutshell: your thoughts are structured like sentences, and the associated thinking is in terms of single lines of thought. How could it be different? Well, your thoughts could be structured like the network we extracted from the brain module which, as we saw, was a fundamentally parallel-processing structure, and which, as we shall see in a later chapter, defies sentence-like interpretation in terms of usefully meaningful local components.

Pinker, for example, (the prime target of Fodor's onslaught[1]) is more positive about roles for neural-networks in models of cognition, but only as components within a LOT-type framework. According to Pinker, "It is the *structuring* of neural networks into programs for manipulating symbols that explains much of human intelligence. In particular, symbol manipulation underlies human language... that's not all of cognition, but it's a lot of it." (p. 112)[13]

As we have seen, the operations of a network of neurons can be interpreted as manipulating symbols. Our information-processing model of the brain module manipulated (by summing and multiplication) symbols that represent link weights and inter-neuron signals. The missing distinction in Pinker's dissatisfaction with neural networks as symbol manipulation devices is that of *level*: neural networks do manipulate symbols, but, insofar as we have been able to understand them, they are not symbols at the cognitive level (another glimpse into the structure-meaning abyss).

So, given that LOT is a contender for specifying the structure of our thoughts — perhaps the sole contender with any reasonable degree of specificity — what is a possible cognitive architecture for our brain module? What sort of information-processing model might we find most useful to aid our journey towards understanding intelligence? The LOT idea suggests that we describe the relevant knowledge and then work with something like the word strings generated. It is well past time for a specific example.

Fortuitously, it turns out that I also know what our small brain module does within the cognitive mind. Recall that I've already let slip what the cognitive level interpretations of four of its inputs are. Here's a reminder: the five input values are the coordinates of two points in a Cartesian space, X1, Y2, and X2, Y1, together with one further value, which we labelled L1.

Now, here's the full picture: the brain module treats the difference between the two X signals (i.e., X2-X1) and between the two Y signals (i.e., Y2-Y1) as the lengths of two sides of a right-angled triangle. The module then determines if the hypotenuse of this triangle is longer than L1 or not. If the triangle's hypotenuse is longer than L1 it outputs a pulse, if not it does nothing. So there's the cognitive-level explanation of our module's processing. How might this processing be represented? A question which I take to be much the same as: what sort of model do we need to extract from our brain-scan data in order to get a start on building a working model, and so begin to develop an understanding of intelligence?

First, let's be clear about the cognitive function of this brain module. After all, a purported "explanation" is only a real "explanation" if you are comfortable with the explanatory components (like "hypotenuse") and the "rules" used to combine them (like "is longer than"). "Explaining," as Dawkins points out, "is a difficult art."[14]

Competence in geometric reasoning is hardly a universal of human cognition — some learn it, some don't, and we all lose it over time if not exercised regularly. (An aside: Is such forgetting indicative of a failure of the brain-machine to do the best that can be expected of intelligence? Or is it a cause for celebration in that it stops our minds from becoming clogged up with information that we have little or no use for? We'll revisit this issue in a later chapter.)

Here's a picture that might well save us all some good number of explanatory words. The five input signals and the single output are illustrated with block arrows.

Points in Cartesian space (i.e., the X-Y plane shown) are designated by an X value and a Y value. So our first point (the black dot upper

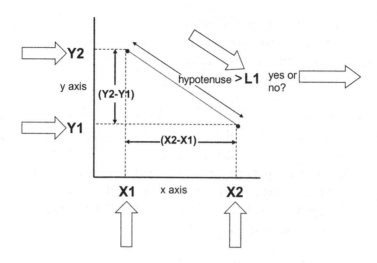

left) is the point at (X1,Y2), and the other point (the lower right black dot) is (X2,Y1). If we draw lines downwards parallel to the Y axis from each of our points, and also lines leftwards parallel to the X axis, we get the four dotted lines illustrated. Two of these lines intersect with a right angle, 90 degrees, the corner of a square. This gives us a right-angled triangle encompassing this corner and the two points.

The "hypotenuse" of this triangle is the fancy name for the line joining the two points (as labelled above).

We can calculate the lengths of the other two sides by means of a simple subtraction of input signals: the vertical side is length (Y2-Y1), and the horizontal side is length (X2-X1). Given the lengths of these two sides, Pythagoras' rule tells how to calculate the length of the hypotenuse: it is the square root of the sum of the squares of the two other sides. In terms of our module's input signals and the mathematical notations for square root and square, the length of the hypotenuse is:

$$\sqrt{(X2 - X1)^2 + (Y2 - Y1)^2}$$

With our competence in geometry re-invigorated, let's move on. (Isn't it fortunate that our various cognitive competences can be re-organised and refreshed so easily? Is it mere fortune?) So what can this brain module be described as doing?

It takes input signal X1 and subtracts it from input signal X2, and squares the result; it also takes input signal Y2 and subtracts input signal Y1 from it, and also squares this result. Then it adds these two results together, and gets the square root of this sum. The resultant value, which is the length of the hypotenuse, is then compared to the value of the input signal L1; if it is greater than L1 the module emits a pulse, if not it does nothing.

From the LOT-based possibilities of how we might think about calculating this little geometric process, we can move on seamlessly to the possibility of a computer program that actually does this calculation. This gives us a possible mechanism for the mind to mirror the operations of the brain module. You do not have to be a programmer to appreciate that constructing a program to compute in the same way as the LOT-based framework outlined above is quite easy. Working with localised, cognitive-level elements of the problem — X1, X2, Y1, Y2, L1 — and with the operations specified by Pythagoras' rule we get:

read in values for X1, X2, Y1, Y2 **and** L1;

if $\sqrt{\{(X2-X1)^2+(Y2-Y1)^2\}}>L1$

then print 1

otherwise print 0

In addition to the LOT-based structure of this computer program, we need to take note of its "flow structure" — the sequencing of the component operations. First the input signal values of X1, X2, Y1, Y2

102

and L1 are "**read**" into the program — i.e., these quantities gain specific values as they must if the computer is to be able to calculate with them. Next all the subtractions, squarings and square-rooting specified by Pythagoras' rule are done to obtain a value to be compared with the value of L1. Finally, dependent upon the outcome of this comparison — i.e., is $\sqrt{(X2-X1)^2 + (Y2-Y1)^2} > L1$ or not? — the program will print out "1" or "0". This flow structure is made explicit in the following illustration.

$$\longrightarrow \text{if} \longrightarrow \sqrt{(X2-X1)^2 + (Y2-Y1)^2} > L1 \longrightarrow \text{then} \longrightarrow \text{signal "1"}$$
$$\searrow \text{else} \longrightarrow \text{signal "0"}$$

In particular, note that there two different *flowpaths* in this program: one leads to the output signal "1" and the other leads to the output signal "0".

The important general point is that every time the program processes a specific set of input signals through to an output signal, a path of the program elements involved in the particular computation can be traced through the program. This thread of "activated" components is the *flowpath* of cognitively-interpretable components from input to output. Usually, this flowpath will be just one of the many that a program embodies; the other flowpaths will be activated by other sets of input values. (Note: This is barely true for this particular semantic model, because the program is so simple. It contains alternative flowpaths only at the very end where the decision is made to output either 1 or 0.)

If we now recast this model of what our mind might be doing when it works through this particular geometric process as a neuron-network-like model, it looks like this:

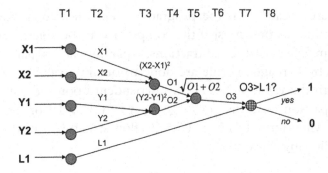

In order to illustrate the complete model I've labelled the three necessary internal signals O1, O2 and O3. Note that flow structure within the Pythagoras rule, which hitherto we've glossed over, is now also made more explicit. Note also that the final node, where the two flowpaths separate, is different from all the other nodes. Every node except this final one gets an input signal, operates on it and outputs the resultant signal by means of its single output connection. This final node involves a test (Is O3>L1?) that determines which of its two output links carries the output signal.

In an effort to better mimic the brain module whose behaviour we are endeavouring to understand, we can collapse these two flowpaths into one that simply carries either a "1" or "0" output signal.

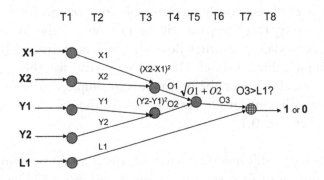

In this cognitive-level model of the behaviour of our brain module I've added some Ts to indicate possible time steps for the module's information processing. In truth, about all we know in this regard is that $(X2-X1)^2$ and $(Y2-Y1)^2$ are independent calculations and can therefore be performed in any order, or even simultaneously which is my choice in the above model. All other steps have a definite order, e.g., the module cannot compare the length of the hypotenuse (labelled O3 in my model) with L1 until it has computed that length — that's just common sense.

We might call this a neuromorphic model. Why? Because it looks like a neuron network. I purposely structured it to look like the brain module whose behaviour it describes. Here's the original brain-module structure for direct comparison.

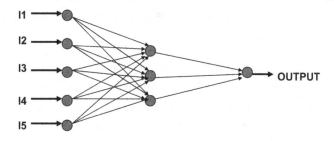

There are clear structural similarities, but that's about the end of the isomorphisms. There are many more links in the brain module. In particular, we note that all inputs signals are transferred to all intermediate processing neurons in the brain module. In our cognitive-level model there is a clear differentiation of the signals in the information flowpaths. For example, input signals X2 and X1 are each transferred to one intermediate processing component, and this component gets only these two signals.

We know also that the links in the brain module are of widely differing strengths. And because they moderate the signals they transmit, similarly widely different signal strengths will be transferred from each input neuron to the three intermediate neurons. In our cognitive model, there is no such moderation of the signals transferred.

In summary, our cognitive model displays a compositionality *at the cognitive level*; the brain module does not. The brain module does display a neat compositionality: each neuron operates independent of the others and operates only on the inputs it gets — it sums its input signals, compares this sum to a threshold value, and then sends on a signal or not.

It is, however, not at all easy to see how to reconcile this brain-structured information-processing model with our knowledge of what this brain module contributes to the intelligence of the brain within which it was found. There is local and systematic compositionality in the information-processing principles of the brain-structured information processing. However, because these desirable characteristics are not at the cognitive level, it does not readily map onto our cognitive-level model.

Maybe we haven't tried hard enough? Perhaps a more systematic and determined attempt to bridge the gap between the brain-structured model abstracted from the brain-scan evidence, and our cognitive-level understanding of the module's behaviour will reveal more? Let's see.

Endnotes

[1] Jerry Fodor's *The Mind Doesn't Work that Way* (MIT Press, 2000) p. 71. This small book (barely 100 pages) was written solely to counter the popular glib accounts of cognition — in particular, Steven Pinker's *The Way the Mind Works* (Norton: NY, 1997 and then by Penguin) and Henry Plotkin's *Evolution in Mind* (Penguin: London, 1997).
[2] W. R. Uttal's "Hypothetical High-Level Cognitive Functions Cannot Be Localized in the Brain: Another Argument for a Revitalized Behaviorism", *The Behavior Analyst* 2004, vol. 27, no. 1, pages 1–6 (Spring), p. 2.
[3] Jerry Fodor's *The Modularity of Mind* (MIT Press, 1983), p. 126.
[4] Jerry Fodor's *The Mind Doesn't Work that Way*, (MIT Press, 2000), p. 88.
[5] Richard Dawkins, *The Blind Watchmaker* was first published in 1986 and by Penguin Books in 1988. Dawkins' "very good reasons for rejecting all such saltationist theories of evolution" (p. 231) do not obviously apply to changes in brain architecture and the consequence changes to minds. One reason being that a small change in brain wiring may cause a large change in mind capabilities, especially if the brain

(like an IT system) is a discrete or digital, rather than analogue, machine (see *The Seductive Computer* for this inherent weakness of discrete systems).

[6] These common illusory images as well as many more and much interesting discussion is to be found in the writings of Richard Gregory, particularly his classic *Eye and the Brain*, Fifth Edition (Oxford University Press, 1997).

[7] Fodor's *The Modularity of Mind*, p. 128.

[8] The subtitle of Kurzweil's book *The Singularity Is Near*, (Duckworth, 2005).

[9] "Every cognitive variable that might indicate a cognitive component reliably interacts with other variables, and the patterns of interaction change across tasks... Perhaps this is why no single cognitive component has yet been discovered for which there is general agreement among investigators." So say G. C. Van Orden & K. R. Paap in "Functional neuroimages fail to discover pieces of mind in the parts of the brain", published in *Philosophy of Science* pages S85–S94, 1997. They lay out the general problem with brain imaging as a route to understanding how the mind works: "One must begin with a 'true' theory of cognition's components, and assume that corresponding functional and anatomical modules exist in the brain. The true theory is necessary to insure that experimental and control images differ by the single component of interest. Additionally, the brain must be composed of feed-forward modules to insure that the component of interest makes no qualitative changes 'upstream' on the shared components of experimental and control tasks. Finally, each contrasted task must invoke the minimum set of components for successful task performance. If any one of these assumptions is false the enterprise fails." (p. S86) And, as their paper makes clear, it is likely that they are *all* false.

[10] W. R. Uttal writing in *The Behavior Analyst* in 2004, vol. 27, no. 1, pages 1–6 (Spring) an article entitled *Hypothetical High-Level Cognitive Functions Cannot Be Localized in the Brain*.

[11] There's a chicken-and-egg issue here. Fortunately, we do not need to address it. The point is not which came first, and hence which way causality flows, but the plausibility of two tightly interwoven systems having similar structuring and basic operations.

[12] Jerry Fodor's *The Language of Thought* (Harvard University Press, 1975), and more recently expounded more succinctly in "Why there still has to be a language of thought", in *The Foundations of Artificial Intelligence: a Sourcebook*, (Cambridge University Press, 1990) edited by D. Partridge & Y. Wilks, pages 289–305 (the opposing "connectionist" viewpoint is put by Paul Smolensky in the subsequent article, and the two classes of theory are discussed in a third article by Yorick Wilks.)

[13] Steven Pinker's hit book *How the Mind Works* (Penguin: 1998) is replete with vague claims (such as the one quoted) which give an impression that we know quite a lot about how the mind works, but boils away to almost nothing once an effort is made to pin down details.

ଔ Chapter 5 ଓ

Meaningful Principles —
The Search Continues

"The purpose of reverse engineering the human brain is not to copy the digestive or other unwieldy processes of biological neurons but rather to understand their key information-processing methods."
Ray Kurzweil, 2005[1]

Our earlier exercise in reverse engineering on the small brain module was perhaps over-optimistic. That was why we ended up perched on the edge of a brain-structured model, staring helplessly into empty space for a route to an unknown cognitive-level model. Wondering where to go as well as how to get there is not a promising basis for progress.

But now, armed with the basic LOT hypothesis and a complete cognitive-level model of what our brain module does, we have, in effect, a hard target across the void. This must improve our chances of constructing the desired bridge from our extracted principles of brain operation to the principles that can transparently account for its intelligent behaviour.

This is the strategy that we'll now pursue in the hope that we can achieve a first traversal (with all conceivable assistance, including the blatant swindle of already having the model we're after). This spanning exercise should generate valuable insights into how we might construct this type of bridge when the details of the target model are not known. Indeed, this cognitive-level model is precisely what the reverse engineer needs to discover at the end of his journey.

To further justify this strategy, I'll invoke the mountaineering analogy (one that gets several more outings in later chapters as the classic illustration of machine learning). But for now, instead of bridging an abyss we'll move this challenge into the vertical plane. The climber in a new and unusual mountain range may usefully ascend her first peak with the aid of ropes affixed to the top. By so doing the thrill (not to mention the point) of summiting may be diminished, but much will be learned about the techniques of climbing in this particular terrain — e.g., what geological features to seek out and grasp or stand on, and which ones to ignore or work around. With such specialised knowledge of feature salience so gained, subsequent, unaided ascents on new, local peaks are more likely to be successful.

In our previous attempt to build across from the evidence of how the brain module operates to an interpretation at the cognitive level, did we make the "toaster error" of following electrons instead of focussing on heating and timing components? Perhaps we did, in effect, "copy the... unwieldy processes of biological neurons" when we should have focussed on trying to "understand their key information-processing methods"[2]. But what is "key"? It is, of course, information-processing methods founded on the *salient features*. Feature salience has already been introduced in the context of the card game Eleusis, further illustrated in the context of toaster models, and cannot (as Maxim 3 asserts) be treated lightly.

Earlier, the non-salience of the "heat-energy" feature of toasters and the missed salience of proximity between heating element and bread were both made apparent because we had two models — a basic structural one and a semantic one — as well as the explicit correspondences, the bridge, between them.

In the quest to understand intelligence, it is this cognitive-level, or semantic, model that we seek. Hence the vicious circularity of the notion of salient features (and hence "key information-processing methods") stumps the reverse engineer who has just one example of intelligence to study: the human one. Once again, toaster systems can illustrate this point of difficulty.

In addition to obvious joints to carve at, toaster systems offer the enquiring scientist another advantage over intelligent minds: just like the skinning of cats, there is patently more than one way to toast a piece of bread. So let's invoke another type of toaster system to demonstrate the gains.

Cognitive scientists have long argued that humanity's long dalliance with hunting and gathering must have left some traces, if not scars, in the modern cognitive mind. Toasting, I venture to suggest, may be just one such vestigial function that we still witness in its most primitive forms. We'll call this campfire process the Boy Scout toaster system, or BS toaster for short. Here's what I have in mind:

As a Boy Scout I learned how to start a fire with nothing more than a large box of matches and some dry paper. I also learned to prong a slice of bread on a cleft stick and then hold it over the campfire until crisply golden (underneath the soot). Nowadays, I slip the bread slices into a chromium-plated box, plug it into the electricity, press down on a lever, and my toast pops up when done. Both systems take sliced bread as input and deliver toasted bread as output.

Neglecting the comparative possibilities for perfect toast, and give or take the likelihood of superficial campsite detritus, these two toasting systems — the BS and the chromium box — are similar in behaviour.

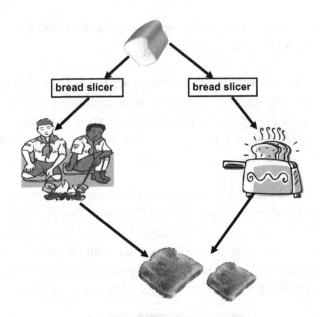

When both are given the same input — two slices of bread — they produce the same output — two pieces of toast.

How does this sample of two different toaster systems help in the search for the salient features of toasting? Here's an example: because the BS toaster requires no electron stream input signal, we know that an electron stream, i.e., electricity, cannot be a salient feature of toasting. However, notice also that they both produce considerable heat-energy output quite separate from that in the toast, so we still have no reason to remove excess heat-energy from the list of (apparently) salient features of toasting.

In general, the existence of more than one type of system that exhibits the behaviour of interest will help rule out non-salient features, but with no guarantees to identify them all.

In addition, the salience of the proximity of bread to heat source, which eluded the scanners of the electric toaster, may well emerge from study of the BS system, but it may not. When trying to understand a system, toaster or mind-brain, the question of what is important to measure and monitor, and what is not, is very difficult to answer — even when we've a couple of different versions of the system to study. It is a cognitive-level understanding of a system that finally sorts all this out.

You and I both know that the heat-energy output feature of our two example toaster systems is a non-salient feature of toasting — i.e., a toaster system does not need to produce this extra heat in order to toast. It is our *understanding* of toasting that tells us this, and it is this sort of understanding of intelligence that we seek!

Hence the vicious circularity of **understanding based on a single example**, which is the problem we face when struggling to understand intelligence with only the human example to study.

Until aliens decide to put their heads above the horizon and allow us to scan them, we are stuck with the single example of an intelligent system — the brain-based one in you and me. With only one example system, science will always struggle to separate the brain's information processing that is essential to its intelligent behaviour, and the information processing that is not. Glib statements about extracting "key features" should get short shrift in any investigation of what makes us clever.

Nevertheless, we did discover the information-processing methods of the neurons. We found that they summed their input signals and compared this total with an internal threshold (the *thresh6*, etc., values). They then output a signal only if the sum was greater than the threshold.

Were these the "key information-processing methods" of the network? Clearly not, if "key" means that the methods reveal useful components for incorporating into a software model of human intelligence. Equally clearly, they were if our goal is to understand the workings of this brain module free of a fixed obligation to end up with the Singularity prize. Like so much else within this ill-conceived notion, more needs to be said about the nature of the desired "key information-processing methods" — in particular, "key" with respect to what?

How do we know what aspects of brain function are required to support intelligent behaviour, and which ones are merely detail that will be superfluous (or even deleterious) to our non-biological model? Such choices need to be made early on in order to set up scanners to monitor and record just the evidence that relates to salient features.

To begin with, we might forego eating, breathing, excreting, etc., on our race to the Singularity, but this helps us not at all in determining which aspects of a biological brain module we can ignore, and which ones we'd better pay close attention to. Failure (such as we've so far experienced) to extract a useful model of our brain module might point to neglect of salient details (such as the role of chemical receptors in a synaptic gap), but it might not.

Salience is ultimately determined by what we need to include in a successful model, and modelling success relies upon a good understanding of salience. This analysis is not looking good; viciousness clearly commends itself as a feature of this tail-chasing exercise. The docile scientist, wisely unwilling to enter this ring, just notes this difficulty and moves on to do the best with what is at hand.

To the envy of every reverse engineer grappling with masses of brain-scan evidence we do happen to have a complete semantic model of the brain module that we've been studying. "Don't ask, don't tell" is the notorious rule that I'm invoking here. Let's just call it a lucky break, and move on. To refresh your grasp of this unwonted coup, here it is:

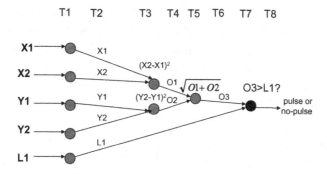

This model is easily programmable to yield a software module for our simulation of intelligence. As a programmable module it boils down to:

if $\sqrt{(X2 - X1)^2 + (Y2 - Y1)^2} > L1$ then 1 otherwise 0

As a result of curious historical accidents[3] this little procedure has long been known as LIC1, and we'll stick with this name.

Is this procedure really a cognitive-level description of our little brain module? If you can't just take my word for it (and when playing the scientist, you shouldn't), then you could test it. How? We can compare its behaviour with that of the model that reproduces the behaviour of the brain module, and see if they are exactly the same.

We first turn the cognitive-level model into a computer program (which will be little more than the bare expression above). For those readers who really want the details, here's the computer program again:

```
read in values for X1, X2,
Y1, Y2 and L1;

if √{(X2-X1)²+(Y2-Y1)²}>L1

then print 1

otherwise print 0
```

To see this program work we give it a set of five input signals, i.e., we give values to the five variables X1, X2, Y1, Y2 and L1, and we record the output signal that our program produces: 1 or 0. By varying the strengths of the five input signals, we can collect input-output pairs, each such pair being five specific input signals and the associated output signal.

Here's what these input-output pairs look like: (1.21, 2.01, 3.22, –0.37, 10.99→1), (–2.22, 9.11, 1.77, 4.12, 7.89→1), (3.45, 8.02, 19.21, 20.33, 0.02→0). They encapsulate the behaviour of the cognitive-level model.

The brain-based model can also be programmed in terms of its information-processing principles — i.e., summing, thresholding and link-weight multiplications. The input-output pairs obtained from the cognitive-level model can then be used to test the programmed version of the brain-based model we extracted from the brain-scan evidence. If the two models really are equivalent, then they should produce the same output signal whenever they are given the same five input signals.

Below I present the brain-scan-based model converted into a computer program for the reader who really wants the details (and who should recall that *w1* to *w18* as well as *thresh6* to *thresh9* are all precise numbers derived from the original brain-scan evidence). Once more, it's not the precise details of this program that are important (you'll no doubt be pleased to know), but the fact that scientists can construct such a programmed model. I present the full gruesome program as no more than an existence proof that such precise models can be built.

If we test these two programs, say, 10,000 times, each time with a different set of the five input values, and on all occasions both programmed models produce the same output signal when given the same five inputs, then all tests have succeeded. This doesn't prove that our two models are totally equivalent in terms of what they compute (no amount of testing can do this because there is always a chance that five previously untested signal values will fail to give the same output signals — recall our logical conundrum, Maxim 6).

```
read in values for I1, I2, I3, I4 and I5;

if (I1×w1)+(I2×w4)+(I3×w7)+(I4×w10)+(I5×w13)>thresh6
then set O1 to1, otherwise set O1 to 0;

if (I1×w2)+(I2×w5)+(I3×w8)+(I4×w11)+(I5×w14)>thresh7
then set O2 to 1, otherwise set O2 to 0;

if (I1×w3)+(I2×w6)+(I3×w9)+(I4×w12)+(I5×w15)>thresh8
then set O3 to 1, otherwise set O3 to 0;

if (O1×w16)+(O2×w17)+(O3×w18) > thresh9

              then print 1
              otherwise print 0
```

However, if the tests are properly conducted, we can get an assurance of total equivalence of behaviour that is a good deal more secure than most (if not all) of the other assumptions we are making elsewhere in this exercise (and indeed that scientists must make all the time).

So what's the claim here? It would be that we have good reason to believe that the cognitive-level model presented above is an accurate simulation of the brain module. This is because when transformed into computer programs, the cognitive-level model, and the model abstracted from the brain-scan evidence behave identically — on 10,000 occasions. For every set of five input signals tested both models generated the same output signal. Such identical behaviour between two models is called *functional equivalence.*

Within the world of toaster systems, we had functional equivalence between the electric toaster and the BS one. With the heater components activated — in one case electron flow into a resistor, and in the other, incandescent combustion of wood — input two slices of bread to each one, and both will output two slices of toast.

The toaster-system equivalences are, of course, a bit looser than our mind-brain ones. I am not claiming that the average Boy Scout consistently pays as much attention to his task as the hardware components of the electric toaster. As a result the quality of the product, the system outputs, are never going to be identical such as we are insisting upon with the input and output numbers of our brain-modelling systems. But the simple idea behind the pompous phrase "functional equivalence" has, I hope, been made clear.

Now, back to more brainy matters. Let's assume the tests have been done and no failures were recorded — i.e., there were no occasions when the two models produced different output signals when given the same five input signals. Consequently, we are happy with the claim of functional equivalence — our brain module is indeed calculating LIC1. They are functionally equivalent systems (at least insofar as the 10,000 tests reveal).

Here's what we have:

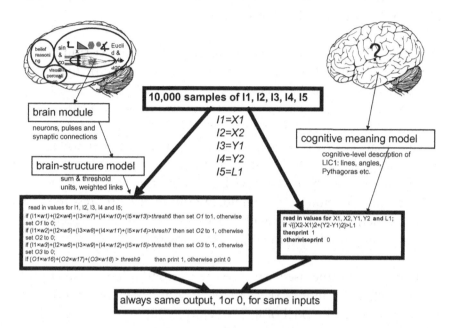

Do not forget that the reverse engineer does not have the cognitive meaning model available. It is precisely this that he or she is endeavouring to extract from the brain-scan evidence. But we are luckier in this particular instance. We hope to capitalise on this good fortune to acquire some valuable insights into how the reverse engineer might proceed in his or her search for a usable semantic model of brain activities.

So now we have two versions of the same system, a brain-structure-based model and a cognitive-level one. It's time to grasp this good news and move on to exploit it.

How can the cognitive-level model be a potential contribution to our task of building a cognitive mind? Well, it shows what one small cog in the brain's geometrical reasoning capability does. It also shows how the brain might be calculating LIC1 in terms of the quantities that you and I consciously use to ponder about such geometrical puzzles — e.g., lengths of lines and triangles, rather than summed and thresholded neuronal signals. We have a cognitive-level explanation that has withstood extensive testing.

Recall that we developed an understanding of the toaster-scan model in terms of our familiar cognitive-level model of an electric toaster by explicitly drawing the interrelationships between the two models. So now, if we can relate this cognitive-level pattern of information processing of LIC1 to the brain-based model (and hence the neuronal module that actually does this processing in the brain), we will surely be making progress along the road to understanding intelligence. Success in relating these two models, building the intervening bridge (or climbing the mountain), should amount to real progress towards a more general understanding of how to reverse engineer the required cognitive-level models from brain-scan evidence.

Can we generate the useful, nay essential, relationships between our brain module, or more likely our brain-scan-based simulation, and our cognitive-level model of it? We can try. Our knowledge of LIC1 has given us a head start over the serious reverse engineer. He or she must discover the pattern-recognition principles that will support, if not deliver, intelligence modelling without the benefit of the sought-after cognitive-level model up his or her sleeve as it were.

But with the luxury of a cognitive-level model at hand, we can approach our goal from the opposite direction. Instead of trying to extract this semantic model from the brain-scan data, we can try to extract the brain-structure model from the semantic one. By means of this ploy, we may be able to complete the necessary connections, and so get some clues about how to accomplish the desired reverse engineering.

To return to the mountaineer — her initial fact-finding exercise should be easier, and perhaps just as informative if, secured in a safety harness, she climbs down rather than up. This climbing down, reverse-mountaineering, will still reveal a lot about the nature of climbing up in this particular terrain. Local knowledge of rock properties and strata conformations so gained should then inform subsequent attempts to climb up similar new peaks. Here's the picture:

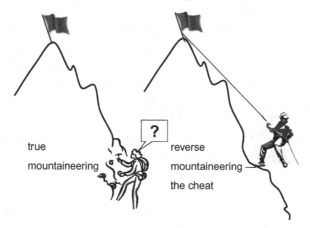

With the proliferation of models and modules, and reversing reverse-engineering, we must be in need of another picture of our particular challenge.

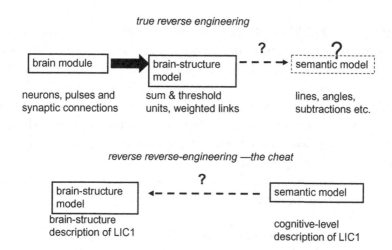

Let's try to work through this easier, "downward" traversal, and so pick up useful tips about how to approach the difficult "upwards" traversal from brain structure to (unknown) cognitive-level meaning, signified as the "semantic model" above.

As we showed earlier, the semantic model does not appear to bear much resemblance to our brain module, nor to the brain-structure model extracted. But because both models are computing exactly the same thing (i.e., they are functionally equivalent), surely we can sort this out? If we can re-jig one or both models to make them look similar, then we should be able to generate an understanding of how the brain module is calculating the various cognitive-level components of LIC1? Consequently, we should then be able to tease out how the brain module is, in fact, computing LIC1, shouldn't we? If successful, we will have a scientific understanding of how one small component of our cleverness is made manifest by our brain.

Just as when presented with two toaster models, say the BS and the electric, if we understand one of them, we can develop an understanding of the other by making the salient relationships explicit, e.g., how the heater component is manifest in each. The hot filament in the electric model does the job of a campfire in the BS model.

Similarly, we should be able to use our understanding of the cognitive-level model to identify the structural elements and information-processing subprocesses within the brain-structured model that do the requisite component calculations of LIC1. Where in the brain-based model is X1 subtracted from X2? Where and how is this difference squared? Where and how is the square root computed? Where is the final comparison with L1 done? Surely these components of LIC1 must be processed somewhere because the brain-based model, and hence the brain module, demonstrably computes LIC1 — as the 10,000 tests have more or less confirmed.

Once we've related the brain module's structure and function to the elements of the LIC1 expression, we'll be a good way along the road to understanding (one small element of our thinking) in terms of brain structure and function. Can the Singularity be far away after such a breakthrough?

How can we begin understanding our brain module in terms of LIC1's cognitive-level components, or something like them? Can we discover some similarities between the models? Can we find connections that at least hint at a role for the processing that goes on within the brain module in the grand scheme of human intelligence?

Before we attempt to climb down from the summit of the cognitive model, we will scrutinise the solid base of the peak, the model based on brain-scan evidence. From this we aim to extract some general guidelines. These, it is hoped, will help us find a promising route down.

Consequently, we begin with an examination of the brain module, the only firm basis for the reverse engineer. It contains four processing elements, the neurons 6 to 9, together with 18 connecting links that also "process" the signals they transfer. This amounts to 22 individual information-processing components with the possibility of subsets combining to produce higher-level information-processing principles. How many such information-processing subsets might we have to consider?

We can assume, very reasonably, that the collaborating processing elements have to be directly connected because other brain cells insulate the inter-neuron connections and so eliminate the possibility of field effects, i.e., interaction in the absence of explicit connections is ruled out. If we had to admit the possibility of a neuron in one part of the brain directly influencing neurons in far distant parts, with no direct interconnections (by means of say, electric field effects), then even our current tenuous hope for understanding disappears. So, we'll rule this out — it's been an implicit assumption from the very outset.

We are still left with many possible subsets of our 22 basic information-processing elements, but not so many as to swamp our current project. So, it might be the case that neurons 6 and 7 work with their 10 incoming links and two output links to process one component of

LIC1, and neuron 8 with its directly connected links processes another.

With these few general guidelines made explicit we are ready to exploit the good fortune of having the cognitive-level model. Let's take a look at how many components of information processing we have on our mountain top. How many such components does this semantic model contain? Perhaps just two: (X2-X1) and (Y2-Y1), with subsequent squaring, adding, square rooting, and comparison with L1 as four operations that need to be performed on our two components in that order. Alternatively, we might want to insist that the supposed component (X2-X1) is just as much an operation (on two of the input signals) as any of the four operations just mentioned. In which case, the model is composed of seven information-processing components: $(X2–X1)$, $(Y2–Y1)$, $(X2–X1)^2$, $(Y2–Y1)^2$, $(X2–X1)^2+(Y2–Y1)^2$, $\sqrt{\{(X2–X1)^2+(Y2–Y1)^2\}}$ and $[\sqrt{\{(X2–X1)^2+(Y2–Y1)^2\}}>L1]$.

In either case, as well as almost any other decomposition you might favour, we seem to have less than about ten information-processing components in the semantic model. This must be good news given that we have 22 individual information-processing components in our brain-structure model with the opportunity to combine subsets to account for each of the few information-processing components that together constitute the semantic model.

Before we probe further into the representational possibilities presented by our brain-structure model, it will be instructive to look at the information flow within the two models.

The brain-structure model, and hence the brain module, receives five input signals, and then the internal information processing proceeds in six steps, where each time step illustrated is designated by "T" with a number attached. Thus "T3" illustrates the information flow and processing at time step 3. Running through the six time steps for the brain module we have:

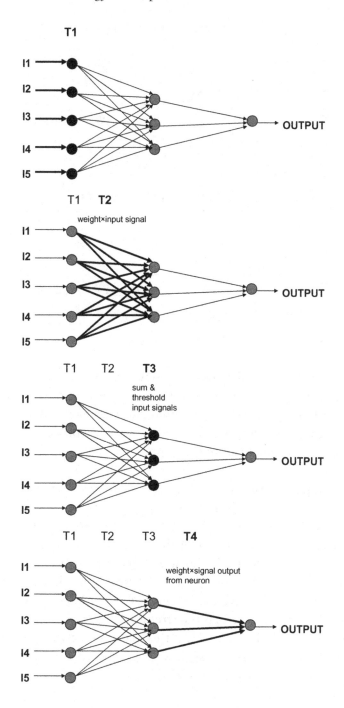

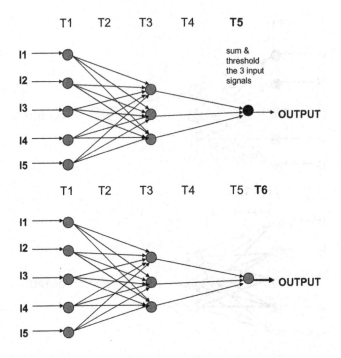

What we see is a wave of signals sweeping through the network from left to right. Every neuron and every link processes a signal every time this network is activated, i.e., at some point in every time sequence (T1 through to T6) in which five input signals are processed through the network to produce an output signal.

If every component of the network participates in every information-processing sequence, we might ask: what makes one such sequence with a particular five inputs signals differ from another with a different five inputs as indeed it must be if this network is to generate results that are input dependent? The answer is that the waves of signals vary in strength across the wave. Each set of five input signals (time T1 above) gives rise to a wave of 15 signals (time T2 above) via links 1 through 15, and what varies is the strengths of the 15 signal elements (one via each link) that constitute this wave — we can call this the *profile* of the wave. A second such wave composed of only three elements occurs at time T4 as illustrated above.

So those are the characteristics of the information flow. Couple this with the information-processing components (also described in the diagrams), and this brain-structure model gives us a pretty complete knowledge of the patterns and principles of information processing in our brain module. With these guidelines to keep us on track, finally we are ready to "climb down" from the semantic model to see how it can be manipulated to directly reflect the brain module's processing — to see how the brain module's processing could account for the semantic model.

Luckily for us it seems that the information flow in the semantic model can easily be interpreted in terms of eight time steps, in a way that looks similar to the brain-module diagrams illustrated above. (In the absence of any guidance from the Singularity literature, I assume that the route to our goal is to make the two models as alike as possible in structure, and hence facilitate the development of interrelationships — hopefully equivalences — between component functions of the two models.)

Here's a run through of the time steps of the semantic model illustrated as a network:

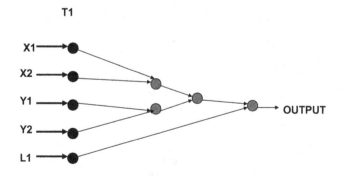

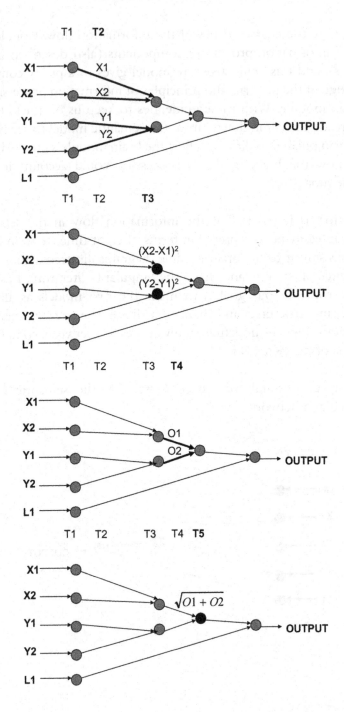

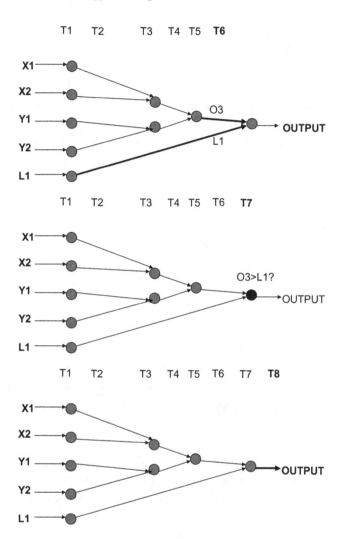

There is a need to add time steps because the nature of the semantic model permits little parallelism — e.g., we must calculate (X2–X1) *before* we can square it. Three intermediate signals have also been added and arbitrarily labelled O1, O2 and O3.

Nevertheless, these information flow-and-processing diagrams do appear to echo those derived directly from the brain module. This is a positive start to our attempt to explicitly interrelate them, and we can make it better yet.

We can, for example, add a processing unit in the link from the L1 input unit to the output unit which will process at time T3, but do nothing more than simply pass on the signal it receives from the input unit. This would give us a diagram (illustrated at time T3) as follows:

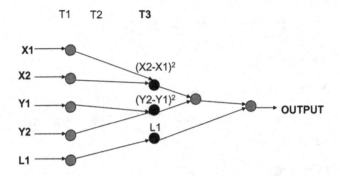

We still have the awkwardness of the extra information-processing steps T4 and T5. But we do have a possible sub-module of information processing to exploit; it is illustrated with an enclosing box below.

Let's eliminate this inconvenience by substituting a single, but more complex, information-processing component. By so doing our diagram really begins to look like a good match to the brain module.

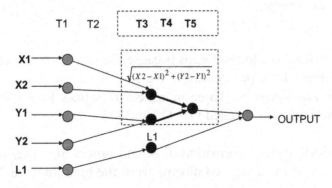

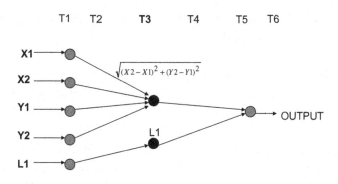

Although we have a good match in terms of both number and layout of information-processing components, the above model is short on links: it contains seven internal links and our brain-structure model contains 18. We can fix this, and at the same time replicate the neuronal function on signal transfer, i.e., what is passed on by a link is *weight × output_signal*. By so doing we get the following representation of the semantic model.

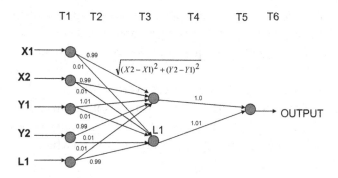

By setting the link weights from, say, the X1 input to the component that effectively computes Pythagoras' rule as close to 1 (I've chosen 0.99), and the link from X1 to the L1-processing component as close to zero (I've chosen 0.01), we get a simulation that looks like our brain module, and yet computes in terms of the semantic-model components. This is because the signal in the first link will be 0.99 × signal X1, which is very close to simply the value of signal X1, and in the other link we get $0.01 \times$ signal X1, which will be close to zero for all (non-huge) X1 signals.[4]

Does all this reversal of reverse engineering mean that now we can be confident that we understand the information processing of the brain module in terms of the cognitive level interpretation of LIC1?

We might propose that two of the neurons chosen from 6, 7 and 8 in effect compute (the specialisation of) Pythagoras' rule, and the remaining one simply passes on the L1 signal to neuron 9. Can this be true? No, is the simple answer. Why?

To begin with, the persuasive isomorphism between our brain module and the above representation of the semantic model is totally superficial. For example, the links in the neuronal module are not weighted in terms of these extremes. The brain-scan evidence shows that it is not the case in the brain that one link from each input neuron carries a strong signal and all the others carry very weak ones. The reality is that various proportions of the links from each input neuron carry a variety of signal strengths; there is no evidence of a simple concentration of the input signal down just one or two subsequent links.

In addition, our simulation assigns very different information-processing activities to each of the three non-input components, whereas the neurons 6, 7 and 8 in the brain module appear to all process information in exactly the same manner (which is not too surprising).

One of the interesting patterns of information processing in the brain module is its dispersal of each input signal across neurons 6, 7 and 8. Whereas all of our semantic-model representations transfer and process the input signals as unitary entities. This latter type of information processing is characteristic of the LOT hypothesis whereas the dispersal of the input signals across all components of the syntactic model suggests the possibility of holistic processing.

So maybe our quest for isomorphism between the brain module, as understood from brain-scan evidence, and its cognitive-level explanation, was misguided. How else might the poor reverse engineer hope to interpret his brain-scan evidence at the cognitive level? I don't know, and the Singularity seekers never give any clues, let alone detailed guidelines.

It's a tough call, probably impossible, even when we have available the cognitive-level understanding that is sought. What hope is there when the reverse engineer, awash with scan data, is groping in the dark for the unknown principles of intelligence to be instantiated in software?

How can close inspection of a brain ever reveal anything not tied to the structure and functions of the anatomical structures found in a brain? How can the beavering nanobots (to allow for the most ambitious scanners imaginable) feed back information above and beyond the information-processing methods of neurons and clusters of neurons?

The semantic model in terms of triangles and lengths of lines might (with a bit of a stretch) contribute to an understanding of how the intelligence of the owner's brain "works". This would make a positive contribution to the brain model that is a major stepping stone en route to our goal of understanding what makes us clever. The functionally-equivalent simulation, in terms of neuronal information-processing principles derived from brain-scan evidence (with the aid of some inspired guesswork), remains an enigma with no obvious contribution to make.

Unlike the mountaineer, who can almost certainly climb down the chosen peak, we failed to connect our semantic model back to the brain-structure one. We discovered nothing new about how to bridge this gulf between what scanning and reverse engineering will deliver, a model based on brain structure and functions, and the

133

cognitive-level models we require if we are to build a scientific understanding of intelligence.

Back to basics: maybe a reliance on the LOT hypothesis as our guide to the architecture of cognition is the big mistake? As I've made plain: not everyone believes in the LOT hypothesis, not even in its basic claims for *compositionality*, etc.; a good claim is far short of an indisputable argument. But initially the agnostics had little to offer beyond their own beliefs that human intelligence had "holistic" aspects that were at odds, in various ways, with the piecemeal *compositionality* exemplified by LOT (and evident in our semantic model of LIC1).

Then in the early 1980s neural computing in a variety of manifestations was promising something new and different. In particular, the *multi-layer perceptron* (or MLP) emerged as a powerful, well-defined and easy-to-use neural computing technology. Ever since, MLPs have dominated the world of neural computing, and well into the new millennium this has not changed. Moreover, MLPs gave modellers a concrete model with holistic properties. Further than this, such a holistic model can be built from brain-scan evidence and yet not be limited to a direct echo of the brain structure from which the scan evidence is collected. It is time to open the curtains and see exactly how this holism is evident in a neural computing model, and whether it might help the reverse engineer to cross the abyss.

Endnotes

[1] A cautionary note from Ray Kurzweil's book *The Singularity is Near* (Duckworth, 2005), p. 444.
[2] Extract from the chapter-heading quotation whose source is detailed in Endnote 1.
[3] Many years ago we embarked on a neural computing exercise to replicate previously published results based on conventional computing. The target problem was composed of a number of Launch Interceptor Conditions (LICs), and the first such condition was LIC1. Over the course of a number of years we generated and explored many neural-network implementations of LIC1. Some representative publications are: (1995) "Engineering Reliable Neural Networks", Proc. 4th *Internat. Conf. on*

Artificial Neural Networks, 26–28th June, Cambridge, pages 352–357, D. Partridge and W. B. Yates; (1996) "Engineering Multiversion Neural-net Systems", *Neural Computation*, vol. 8, no. 4, pages 869–893, D. Partridge and W. B. Yates; (1996) "Network Generalization Differences Quantified", *Neural Networks*, vol 9, no. 2, pages 263–271, D. Partridge. **Note**: all of the detailed networks in the illustrations in this book have been dreamed up, i.e., the actual link weights are fiction, so they will not perform as claimed (except by chance). However, the reality of the claims is true. The doubting reader can find the details on how to construct working examples, and how they performed, in the above-referenced papers.

[4] This piece of mathematical trickery is simply capitalising on the fact that any number multiplied by 1.0 (or a value close to 1.0) will be (almost) unchanged, and any value multiplied by zero (or a value close to zero) will become (almost) zero.

෬ Chapter 6 ෭

Holism — an Unholy Problem

"There is a great difference between mind and body, in that body, by its nature is always divisible and that mind is entirely indivisible."
Descartes, 17ᵗʰ Cent.[1]

"The result [of early AI endeavours] was an account of central processes which failed to capture precisely what is most interesting about them: their wholism... What emerged was a picture of the mind that looked rather embarrassingly like a Sears catalogue."
Jerry Fodor, 1983[2]

We've seen how our toaster models variously interrelate via the salient features of toasting, and that through such explicit interrelationships a more complete understanding of toasting develops. We've also seen that attempts to similarly interrelate "brain" models, and so understand a component of intelligence, have yielded no success. This is because we don't know what the salient features of intelligence are, or if we do, then we have no idea how these features, such as memory, are realised in terms of brain structures and behaviours.

After all our efforts have come to naught, are we any wiser?

The previous chapter concluded with the observation that a major point of mismatch between brain processes and a traditional description of intelligent behaviour, a semantic model, was the seeming incompatibility between the fundamental information processing of the two models. The brain module, and hence the evidence-based model extracted, consistently distributes all internal signals. In contrast, all information processing within the semantic model rigorously maintained the input signals as localised packets of information, or transformed them by way of cognitively meaningful operations into further localised packets of information.

For example, in the cognitive model we could trace, say, the input signal X1 and its derivatives, (X2-X1), then $(X2-X1)^2$, and so on through the model. At each stage it was possible to explain the packets of information being processed as well as the purpose of the operation being applied, and these explanations are at the cognitive level — i.e., they constitute the "understanding" that we seek.

Rather differently in the evidence-based model, which necessarily echoes the structure of the brain module from which the evidence was derived, the input signal X1 is immediately dispersed unequally among the three internal processing units. The other four input signals are similarly dispersed. So right from the outset we are denied a straightforward cognitive-level explanation of the information processing, and it just gets worse with (cognitively) meaningless summing-and-thresholding everywhere followed by further dispersal.

A summary of the mismatch between localised and distributed, or cognitively meaningful and holistic, information processing within our models is:

1. The localised semantic model processed and transferred individual "packets" of information through one of the flowpaths of the model. (Our simple model contained only two flowpaths, but in general a semantic model may contain many different flowpaths.) These packets of information corresponded to meaningful elements of the problem at the cognitive level, e.g., coordinates of a point as X2 and Y1. Each information-processing element of every flowpath also corresponded to a similarly meaningful process, e.g., if length of the hypotenuse >L1 then output a pulse.
2. The distributed model, which we reverse engineered from brain-scan evidence, necessarily copied the information processing observed in the brain module. It distributed the five "packets" of

input information amongst all its processing units. The model contained *no* distinct *alternative* flowpaths. Every link and processing unit takes part in every computation the model performs. It is the profile of a wave of information that characterises each different activation, not a flowpath traversed. We did not manage to relate components of the information processed and transferred with any cognitive-level elements of the problem.

This distributed information processing evident within the brain module was interpreted as the first hint of a holistic process. Hitherto, practitioners in all walks of life from mental well-being and natural-health studies to the efficacy of alternative medicines have tended to reach for holism as an explanation when all others fail. Or when they simply want to distance their practices from the supposedly cold objectivity of empirical science. The appeal to "holism" is all too often short-hand for: we don't know how this works or why, but it does (or we're claiming it does)[3].

Hardly the stuff of science, one fervently hopes. In our quest for a scientific understanding of intelligence, the objectivity of evidence-driven enquiry is exactly what we want. Holistic information processing does not preclude it despite the implication of unfathomability so often, and so cheerily, embraced by advocates of holistic processes.

We'll see holism variously manifest as a variety of different, but all well-defined, principles of information processing. We'll also see that holistic systems can be both constructed in accordance with, and understood in terms of, normal scientific practice. Such a demystification of holistic systems is designed to demonstrate that the scientist need not necessarily throw in the towel of empirical objectivity just because the brain (or even the mind) turns out to be significantly holistic in nature.

To focus on specifics we'll dip once more into the rich universe of toaster systems to illustrate the basic contention that even when faced with a holistic information-processing system, all hope of relating it to a localised cognitive-level model is not lost. Such a set of interrelationships would then amount to an understanding of a holistic process in terms of the (seemingly) necessary localised one. Can we achieve this?

Consider the Cosmic Toaster. When contact is finally made with other intelligent life forms, many new and mind boggling systems will no doubt come to light (not to mention an invaluable second example of "the intelligence program"). Within the plethora of wonderful new systems the Cosmic Toaster is one sure bet.

Although these visiting examples of novel intelligent systems may exhibit an understandable reluctance to let Earth scientists delve into their organs of cognitive excellence, they may well be more open with their state-of-the-art toaster — once assured of commercial confidentiality and acceptance of inter-galactic licensing agreements. On the assumption that the multi-language instructions that accompany the Cosmic Toaster, even if they include an English version, are likely to be no more informative than the terrestrial ones we normally grapple with, the Earth scientists will have to probe and experiment in order to learn how it works.

It will be spherical object (I'm guessing) with all the mysterious promise of a football. How does it work? Simple. You push a loaf of bread into it, and perfectly browned slices of warm, crisp toast emerge from the opposite side. Yet the toaster itself remains cool although there is a suggestion of a little steam leaking from some of the seams.

Anyway, here it is, and it clearly exposes the parochial nature of the terrestrial adage that making toast can be tricky*

---

*"I was quite grown up before I learned that you didn't have to make toast by burning the bread and then scraping off the black bits." So wrote Lorna Sage, commenting on her mother's poor parenting skills in her biographical *Bad Blood* (Fourth Estate, London: 2000). In our language, time and repetition had disabused Ms Sage of the salience of overcarbonising-the-slice-surface as a feature of toasting.

As a toaster, it is functionally equivalent to an electric toaster, or a BS toaster: the input is bread, although not pre-sliced, and the output is toast.

Here is our diagram illustrating functional equivalence between different toaster systems further extended with the newly revealed Cosmic Toaster.

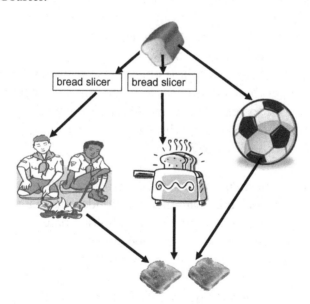

But that's where the similarities end because the Cosmic Toaster will, of course, be a holistic toaster. When our scientists open up this mysterious sphere they will discover only a seething mass of nanobots — no heater element, no timer, no pop-up wire basket, no diminutive

Boy Scouts with sharp sticks at the ready, no internal structure at all (above the level of the nanobot). It is a distributed information-processing system seemingly devoid of localised, cognitive-level toaster components.

How does it really work? A good question and one with which Earth science will struggle. However, its holistic fundamentals do not necessarily condemn the hungry scientist with a penchant for the latest gadgets to the frustration of toast making in complete ignorance of how it's done. So how then is it done?

As soon as a loaf of bread is input, a large number of minuscule nano-engineers instantly and (probably) individually set about restructuring the bread molecules in layers whose width is determined by the toast-maker thinking hard about "thickness". Inter-slice carbohydrate and gluten links are severed, intertwined molecules are prised apart, oxygen and hydrogen atoms are stripped out and ejected as water, and many other water molecules are whisked away from the carbon-enriched emerging surfaces. Each of the nanobots executes one, some, or perhaps all, of these operations. At the same time some, or possibly all, of these bots appear to be designed to operate particularly roughly and so agitate the molecules they are working on as well as all surrounding molecules, otherwise it is hard to explain why the toast is not cold on emergence[4].

This is, after all, nothing but a toaster, a fancy one I'll grant, but only a toaster (despite obvious potential for secondary use in a kick-about after breakfast). With this insight into its holistic nature, can we now relate its components, structures and operations, to our newly enriched, cognitive-level knowledge of toaster systems?

No, it doesn't have any internal components. Well it does really, doesn't it? It is stuffed full of individual nanobots that each perform one or more of several well-defined operations — severing intra-molecular links, hustling water molecules out of the bread, elbowing intertwined molecules apart, ripping out hydrogen and oxygen atoms, and generally shaking up all the atoms and molecules that they come across.

With a little more study and some advances in chemistry and physics at the molecular level, complete and precise explanations of the patterns of information processing within the Cosmic Toaster are entirely possible. But they will be explanations of the information processing at the atomic level. Viewed from the cognitive level this toaster is unswervingly holistic.

Suppose we could, for example, suck out (or otherwise disable) a random selection of half the nanobots, would the toaster cease to toast? Probably not. The likely repercussion of this drastic intervention is that it would toast more slowly with perhaps a noticeably inferior product — less well and evenly browned, and perhaps not piping hot. Almost any such random intervention with the innards of the electric toaster, let alone removal of half its insides, will likely result in a non-functioning system.

Robustness of behaviour is a general characteristic of holistic systems, one that sets them apart from their localised equivalents[5]. This distinction is also a further pointer to the possibility of holistic features of the mind-brain complex, but like so much else that relates to intelligence, the pointing is neither precise nor simple: humans have an amazing capacity to regenerate aspects of behaviour after destructive trauma such as strokes. However, which behavioural capabilities survive brain damage, which return and which do not, is ill-understood. Younger brains appear to be more resilient in this respect, and that is not at all surprising. Ultimately, we cannot say much more than that, in general, the evidence of the recovery from brain damage (not to mention resilience to continual neuron death[6]), with regard to behaviours such as memory, is in itself a forceful argument for distributed, aka holistic, information processing.

A more direct indication of holism in the wondrous new kitchen appliance is, for example, the absence of a heating component as such in the Cosmic Toaster. The heating appears to be a by-product of some (or all) of the rough stuff required to sever inter-atomic links in the "bread" molecules — whether it's part and parcel of "slice development" or due to the ripping off of hydrogen and oxygen atoms to generate a carbon-enriched slice surface, or both, may be fundamentally indeterminate[7].

Nevertheless, there is a real sense in which we can expect to be able to understand how the heating component, a presumed-salient feature of toasting systems, is realised within the Cosmic model. It is distributed among all (or some) of the nanobots, and in either case random removal of half the component nanobots is unlikely to result in the production of stone-cold toast. The loss of optimal heating is likely to be accompanied by a similar loss in all other salient features of toasting: slice surface carbonisation, and in the special case of our Cosmic Toaster, slice separation could well be incomplete.

Alternatively, if the Cosmic Toaster has been properly designed by a distinctly higher form of intelligent life, then the half-disabled version will just take longer to produce top-quality toast. This is because the remaining nanobots will most probably retain the full scope of behaviours necessary to produce good toast. There will just be less of them and internal quality controls will ensure that all the essential characteristics of good toast are met before a slice is allowed to emerge. With half the number of nanobots to do the job we can expect toasting to take about twice as long.

Notice that this new example of a toaster system reveals an earlier misconception about the salient features of toasting: proximal heating is not (as we had previously concluded) a salient feature of toasting. Contrary to both the electric and the BS toaster systems, the Cosmic Toaster does not use proximal heating to produce the carbon-rich surface of a slice of toast. The Cosmic version achieves this necessary characteristic of toast by explicitly severing inter-atomic links in order to remove the hydrogen and oxygen atoms from the superficial carbohydrates.

As a result of our new example system we can replace our earlier conclusion that a "heating source to brown bread" is a salient feature of toasting with, say, "carbon enrichment of slice surface". This salient feature just happens to be achieved by proximal heating in most (maybe all?) terrestrial toaster systems. Hence our mistake. It is the appearance of a new example, the Cosmic Toaster, that permits this refinement of our understanding of toasting.

Holistic systems present new challenges to understanding a complex system in terms of more or less independent components, because there aren't any at the cognitive level. That's what "holistic" means. But, as we have seen, with the Cosmic Toaster, lower-level components may permit some sort of higher-level, localised understanding. We can, for example, make some sense of the heater component of the Cosmic Toaster in terms of nanobots agitating inter-atomic links, which (the science of chemistry tells us) generally results in a rise in temperature. And there is the real possibility that scientific study could pin down as precisely as desired exactly how the Cosmic Toaster produces toast that is hot. By this means, the cognitive-level concept of the necessity for a "heating" component in all toasters may be realised in terms of a comprehensive, atomic-level explanation in the Cosmic Toaster.

So, all hope of an understanding in terms of everyday localised semantics is not lost when confronted with a holistic system. It all depends on the nature of the holism, and how much we know of the relationships between the higher-level and lower-level operations. Earth scientists know, for example, quite a lot about the relationship between heat and inter-atomic energy levels.

Consequently, crossing the abyss from the holistic Cosmic Toaster to a cognitive-level understanding such as we have in the electric toaster does appear doable. In other words, there appears to be no insurmountable reason why we cannot develop a reasonable understanding of the holistic Cosmic Toaster, despite the fact that our understanding will need to be rooted in our localised cognitive-level concepts of everyday toasting.

An understanding of the workings of the Cosmic Toaster (indeed, recognition that it is a toaster and not just an intriguing example of a truly universal constant associated with the leisure-time pursuits of all intelligent life forms) can be gained by relating its structure and operations to those of our cognitive-level understanding of toasting as conveniently exemplified by the household electric toaster. It is the salient features of toasting that anchor the bridging relationships.

Here's a picture of our first steps towards constructing the necessary bridge of interrelationships — a localised understanding of a holistic system.

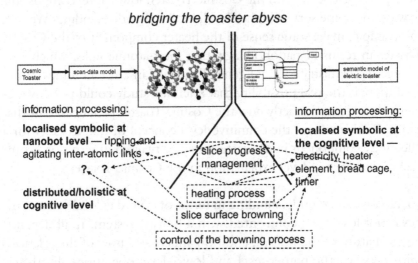

bridging the toaster abyss

Fortunately, human brains appear to be less uniformly holistic than the Cosmic Toaster. It is evident that, anatomically, the brain is composed of distinguishable "regions" that are structured and function noticeably differently, but whether and to what degree the architecture of cognition supervenes on brain structure is anybody's guess[8].

The reverse engineer may be reassured by the discovery of brain regions, but the task is explicitly to construct a mind, not a brain. So the reassurance may be illusory, and who knows to what degree? Just another manifestation of the structure-meaning divide — form does

146

not necessarily mirror content, especially in a non-systematically-engineered construct. In his book *Kludge: The Haphazard Construction of the Human Mind*[9], Gary Marcus reviews the limitations of our cognitive architecture. He concludes that it's a bit of a mess which is to be expected in a product of evolution, although Richard Dawkins (in a view previously aired) might be expected to disagree[10].

Regardless of expectation with respect to the intelligent mind, we do have available two different models of the same process, and yet we have found no way to interrelate the components of one with those of the other. It is as if the brain-scan model is constructed along the holistic lines of our Cosmic Toaster, and the readily understandable, semantic, model of the LIC1 process echoes the structuring of the electric toaster, yet our attempted bridge-building failed utterly. Why?

What we are missing is the science that ties neuron-level information processing to that of a cognitive-level understanding of intelligence; we have few clues about the necessary intervening concepts, the salient features such as a "heating process" in the concept of toasting.

However, toasting is an easy example, and not just because it is arguably a far simpler process than thinking. Toasting is easy because we have a lot of solid information about both systems: the Cosmic Toaster and the electric one. When we switch to trying to understand our brain module, we have no such solid grasp of the semantic model. Indeed, the whole purpose of scanning the brain module is to extract a semantic model that can be programmed and built into a simulation of intelligence, an AI system, and so permit detailed exploration and understanding.

Here's a picture that summarises our problem. On the left side is our original brain module and the syntactic model that we derived (via scan evidence) from it. On the right side is our semantic model, the cognitive-level understanding of LIC1.

the nature of the intelligence abyss

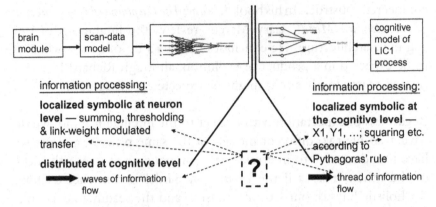

It is difficult to imagine how these two fundamentally different models could ever be reconciled as equivalent models of the same phenomenon because we lack any inter-relationships between neuron-level processes and cognitive-level ones. In toaster terms we lack, for example, the knowledge that links inter-atomic energy levels to heat generated in slices of bread.

What are the equivalents of "heating process" and "control of the browning process" within intelligence? What are the salient bridging features of intelligence? What might go into the central dotted box to replace the big question mark?

Concepts like "memory" and "learning" might be good candidates because the fact that we can accept new information, and preserve it over time for subsequent application — i.e., learn it and memorise it — appears to be crucial to intelligence. Encouragingly, much of our cognitive-level understanding of LIC1 has clear "memory" elements that must have been "learned" — Pythagoras' rule as well as basic geometric concepts such as triangle. But the scan-data model, as well as the neural network from which it was derived, does not present any obvious characteristics of "memory".

Notice also that an absence of memory-like characteristics might be expected as a consequence of one of our initial stipulations. It's a restriction that facilitates interpretation of the brain-scan evidence. Why?

Recall that, apart from its five input signals, the brain module we studied had no connections into it from other parts of the brain. It was informationally encapsulated. This stipulation ensured that no changes were introduced in the course of its multiple activations (the 10,000 experiments necessary to generate the evidence). We explicitly excluded learning behaviour. We shall see below that the shift to holistic modelling opens up the possibility of learning a distributed memory.

The option of holistic information processing in the brain is one that the Singularity programme explicitly promotes (whereas neither LOT, nor its basic characteristics, merits even a dismissive mention). Some "new", nature-inspired technologies, such as neural computing, present us with an automated procedure for building holistic models from minimal brain-scan evidence. We can now construct holistic computational systems, although not yet holistic toasters (for Earthlings early in the second millennium — if it looks like a football, it most probably is a football).

It's true that our initial information-processing model hand-crafted from nanobot brain-scan evidence was holistic. So what's the gain to be realised by focussing on the technologies of holistic-model building?

Firstly, the light-touch nanobots that generated the evidence of neuron signals into, inside, and out-of each neuron do not exist. Despite the flurry of research and development of nano-technology, they are not close to imminent existence. Hence, we cannot in reality collect such detailed evidence as I hope the chapter on brain-scanning technologies made clear.

The good news is that holistic-model building technologies do not require this unobtainable level of evidence detail. They require only a large number of examples of five input signals each associated with the corresponding output signal produced by the brain module — input-output pairs, you may recall. Using the "invasive" brain-scan technology of single-cell recording, it is just about plausible that such input-output pair evidence could be collected from our brain module under study.

Secondly, the process of deriving the model from the empirical brain-scan data is fraught with problems. We used it to exemplify the scientific process: examine evidence, frame hypotheses, and then test with more evidence. Our use of this process proceeded very smoothly. All our initial hypotheses were subsequently supported by further testing, and thus we accepted them all and built a model based upon them.

The reality is that the scientist would be confronted with a mountain of evidence, thousands and thousands of neuron-signal values, and the possible hypotheses are legion. Added to this, the evidence will never be clean and simple. Neurons are not machines, and so they will not exhibit the simple repeatability of machines. Hence, testing will never be clear-cut. The complexities of collecting the evidence, which includes dealing with brains that are always going to be involved in other activities (such as keeping the associated body alive) as well as technical malfunctions of the measuring processes, will inevitably add error to the relevant signals and even lead to totally spurious signals.

The overall consequence of this is that any models extracted will contain elements of nothing more solid than inspired guesswork. This means that certain structures and processes within the derived model will rest on nothing more substantial than hunches and hopes that do not fly too squarely in the face of too much evidence. The holistic modelling technology we will use offers better guarantees from a vastly simplified sequence of actions.

Thirdly, if we eschew all evidence of internal signals, then the resultant modelling procedure is liberated from the constraints of the brain module's internal structure — it must be, mustn't it? Consequently, our new holistic model will not be just a slavish copy of the brain module's structure. By so avoiding these structural constraints we may just open up a possible route around the structure-to-meaning abyss.

Consequent to these points in favour of holistic-model building we'll take a closer look into an appropriate technology (and at the same time take a first dip into one of the "nature-inspired" computing technologies that some believe promise so much). I refer to "neural computing" — the construction of computational models composed of so-called "artificial neural networks" which are no more than computer programs.

It is, in fact, well-known and proclaimed in one of the general names for neural computing — Parallel Distributed Processing — that these computational technologies do disperse the information they process throughout the whole network, i.e., they are fundamentally holistic systems.

A traditional (i.e., LOT-based) computer-program model of LIC1, as we've seen, carries the five input values through sequences of information processing such that all elements of the process have a straightforward interpretation at the cognitive level. This is a necessity for the simple reason that some human programmer has to construct this program, and can only do this by combining elements that have a meaning at the cognitive level, and combining them in ways that are also understood at the cognitive level.

It will not be possible to construct a holistic model that does not slavishly echo the structure being modelled by such classical programming techniques. Classical programming is founded on the combination of cognitively meaningful components and processes as localised elements of the program. A holistic model, almost by definition, is not primarily composed of localised, cognitive-level elements.

So how do we construct a holistic model and yet avoid the constraints imposed by the structural details of the brain module to be modelled? The answer is, the model is automatically "learned" from the external behaviour of the brain module — i.e., the signals that the module accepts and the signals it consequently generates, a set of input-output pairs. This smacks of the magical, so let's look at some details in order to eliminate this unwanted (and unjustified) sensation.

As introduced at the conclusion of the previous chapter, it turns out that the well-founded and most widely used neural computing technology, called MultiLayerPerceptrons (MLP)[11], generates models that exhibit holistic properties. Even better, it also turns out that our earlier brain-scan-based model of the brain module is almost an MLP, i.e., its basic structure and processes (summing and weight × signal link transfers are two of the component processes of an MLP). The "almost" reservation refers to the necessity to replace the hypothesised threshold operation with one known as "squashing"[12]. All that this squashing amounts to is that the 1 or 0 output signals that we derived from brain-scan evidence become replaced by values between 1 and 0 in an MLP model. Each output signal, however large or small it turns out to be, is thus said to be "squashed" to a value between 1 and 0, never being quite as large as 1 or quite as small as 0.

Why this odd operation? For no better reason than that it is required by the mathematics that guarantees that the learning process for MLPs will always be a process of improvement, or no change. The MLP learning process will never cause the model's behaviour to become worse. This is a very valuable property of a machine-learning process, but not, of course, a property of most human learning processes.

When reverse engineering the brain module in Chapter 2, we posited hard thresholds as a result of rounding the neuron output signals

152

recorded[13]. This presumption may be a correct characterisation of the brain module's information processing, but for an MLP it is incorrect. The MLP technology stipulates the necessity of this squashing process. But, because we require either a "1" or "0" as the final output signal from our model, we do "threshold" the final squashed MLP output to transform it into "1" or "0".

So the holistic model that we aim to generate, the MLP, will exhibit similarities to our original brain-scan evidence based model, but also some significant differences. How then do we use automatic learning to generate the MLP model?

We collect the evidence of, say, 1,000 different input-output pairs, i.e., 1,000 sets of five input signals each paired with the corresponding output signal that the brain module generated. We can then use this evidence together with the MLP technology to "learn" a holistic model of the brain module. Here's one MLP model that might be learned:

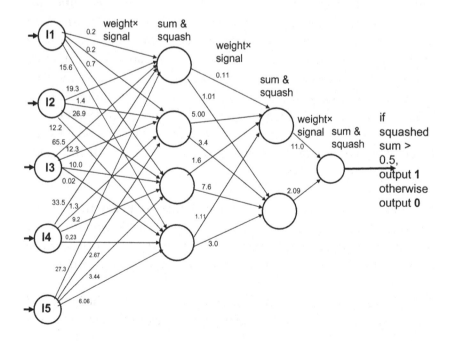

This model should look comfortingly like the syntactic model that we deduced from the full nanobot-derived evidence in Chapter 2. It has a few more internal processing units but other than that I hope it's familiar. All well and good, but how did we get this MLP model? Where, for example, did all the link-weight values (all 30 of them) come from?

The answer is: through a machine learning algorithm. An appropriate set of values was automatically *learned*. The MLP-learning algorithm calculated and set these 30 values for us. How did it do this?

Recall we have 1,000 input-output pairs — five input signals each coupled with the; correct output signal — which were collected from experimental activations of the brain module. Each such "pair" is five possible input signals each associated with the output signal that the brain module produced, i.e., the correct output signal for these particular input signals.

Each set of five input signals is fed into a "randomised" MLP model and the corresponding output signal is recorded. The difference between the final squashed-sum output signal generated by the MLP model and the correct output signal provides an error value that the MLP machine-learning procedure uses to automatically adjust all the link weights. This weight adjustment procedure is repeated hundreds or perhaps thousands of times until all 1,000 of the MLP's output signals are correct. In which case, the MLP has learned to reproduce the brain module's behaviour. Such a "trained MLP" model is then functionally equivalent to the brain module from which the input-output pairs were collected

By "randomised" MLP model, I mean that to begin the process, random values are given to the 30 link weights. By "learning", I mean the automatic adjustment of these weights until the MLP computes the correct output for each of the evidence pairs, the training samples.

The classic view of machine learning procedures (which we will develop further in a later chapter) portrays learning as a process of slogging uphill in an effort to reach the best version, which is on the hilltop. In these terms, here's the picture of the training of our MLP model to correctly reproduce the behaviour of the brain module.

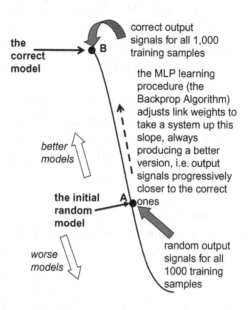

the correct model → B — correct output signals for all 1,000 training samples

the MLP learning procedure (the Backprop Algorithm) adjusts link weights to take a system up this slope, always producing a better version, i.e. output signals progressively closer to the correct ones

better models

the initial random model — A

worse models

random output signals for all 1000 training samples

We started with a model at some random point on the slope (illustrated as point **A**), and we used the MLP learning procedure to progress steadily up the slope to the top (point **B**). This is the top because our model can do no better than correctly reproduce the brain module's behaviour as captured in our evidence of 1,000 input and output signals. The learning procedure was nothing more complicated than gradual adjustment of all the link weights — the 30 such weights in our particular example illustrated above.

How could you, the human programmer, make the necessary adjustments to the 30 link weights? Of course, you couldn't, but crucial to this MLP technology is an algorithm that will make these adjustments for us, and its use is a vital part of building this model[14]. It is precisely the existence of this well-founded machine-learning[15] algorithm that makes MLP technology so valuable.

Other aspects of this automated model building involve choosing the number of internal processing units (I chose 6), how they are organised (I chose two "layers", one of 4 units and a second of 2 units), and the exact squash function to use[16]. My choices were demonstrably okay because my network successfully trained to become functionally equivalent to the brain module from which the evidence was collected (in a subsequent chapter on these "new" technologies we'll delve a little deeper).

In this manner we get a new working model of our brain module, one that exhibits holistic properties but is not tied to the structure of what it is modelling. The brain module contained 3 internal neurons and 18 inter-neuron links whereas our MLP model contains 6 internal processing units and 30 inter-connections.

So what determines the structure of the MLP model? The input units must be 5 with a single output unit. Obviously, these two aspects of the structure are fixed by the brain module that we are modelling — it accepts 5 input signals and generates a single output signal. But what happens in between is up for grabs, within limits. The limits are determined by the complexity of the behaviour of the brain module, the more complex the behaviours to be modelled, the more internal structure will be required. Because these limits are loose, the model builder has a good deal of choice: in this case between perhaps 2 internal processing units and maybe 8, and between, say, an organisation of 1, 2 or 3 internal layers. The model builder also selects a specific squashing process, and then finally connects each internal processing unit to all others in the next layer. These connecting links are then given random weights, and the network is ready to learn to reproduce the brain module's behaviour.

This learning, which is done automatically, amounts to small-step-by-small-step continual adjustment of the link weights until the network correctly computes the set of input-output pairs. "Correctly computes" means that when the model is given 5 input signals, it generates the same output signal as the brain module did when activated with the same 5 input signals. The learned model is thus functionally equivalent to the brain module. But, in contrast to our original brain-scan evidence model, it is not constrained by the internal structure of the brain module, nor did its generation require any evidence of internal signals.

Notice incidentally that we now have a candidate for "memory" (a possible salient feature of intelligence) in our holistic model. The randomised initial network knew nothing and the final trained version can compute LIC1. So where is its "memory" of how to compute LIC1? It must be the set of link weights because they are the only changes that occurred during learning. Is this set of 30 weights (and their detailed interrelationships) in effect a distributed memory of the geometry needed to compute this function?

The MLP model is fundamentally holistic. What does this mean, and how can we be sure? In order to explain the holistic features of our MLP, let's look at the information processing within it (as we did for the traditionally programmed model).

The MLP model we have is[17]:

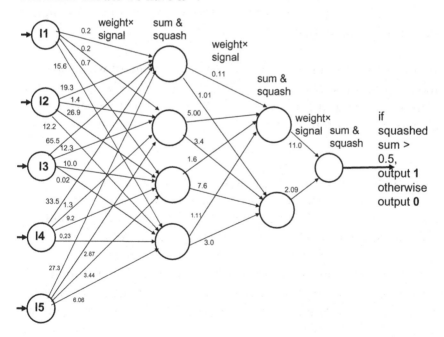

As with our earlier brain-scan-based model, this network also works in terms of parallel waves of information. In this case one wave from the five input units to the first hidden layer of four processing units, and then a second wave to the hidden layer of two processing units, with finally a third wave to the single output unit. This MLP model thus processes information in terms of waves of signals sweeping through the network, and different computations (each triggered by a different set of input signals) is manifest as differing wave profiles — i.e., different sets of signal values.

Consequently, there is no flow structure composed of alternative flowpaths, or threads of execution. Every link and every processing unit takes part in every computation. All this should be familiar: it is an echo of the description of the initial model manually extracted from the full brain-scan evidence.

A further, and less welcome, consequence is that the internal information processing of this holistic model is not looking promising as a basis for pursuing our quest for a cognitive-level understanding of how the brain module works, even though it is no longer simply a slavish reflection of brain-module structure.

Insider information has revealed the understanding of what the brain module does as succinctly defined in the process we've called LIC1. This means that the holistic essence of our MLP model can be clarified in relation to its non-holistic equivalent, namely LIC1. Let's do that.

Because the MLP model reproduces the behaviours of the brain module (this is what the machine learning achieved), it must also correctly compute LIC1 (i.e., when given five input signals, it outputs the same output signal as the traditional program does).

This being the case, we have licence to assume that it must be, in effect, calculating the lengths of two sides of a triangle and applying Pythagoras' rule. Somewhere, somehow, this MLP is subtracting X2 from X1 and squaring the result because it is computing LIC1. But immediately on input, X2 and X1 are both dispersed to all four internal processing units that comprise the first internal layer. These four units also receive input signals from the other three inputs at the same time, and so all hope of localising the subtraction operation is gone.

It should, then, be crystal clear why we cannot match elements of our conception of this computation with any components of the MLP-based computation. The MLP is computing the LIC1 function *as a whole* with regard to our cognitive-level model.

The MLP contains clear-cut computational components, e.g., each of the internal processing units independently sums and squashes the input signals it receives. This is simple and clear symbol manipulation, but not at the cognitive level. *The holistic nature of this information processing is level dependent*, and once more highlights the importance of this distinction (one I have been labouring in terms of *structural* versus *semantic* models).

Recall that we generated a reasonable expectation that a holistic Cosmic Toaster could one day be understood in terms of the more conceptually manageable localised toaster model. But with MLP-type holistic information processing, this anticipated development seems to vanish. Why is this?

It is because the MLP model is composed of components and operations that do not relate in any obvious way to the localised components of the LIC1 process. For example, what has summing and squashing got to do with lengths of sides of triangles, or with the components of Pythagoras' rule? We do not know. In other words: we lack knowledge of the salient features of intelligence, features needed to bridge across the abyss.

Within the Cosmic Toaster, by way of contrast, we know that the energising of intra-molecular links will cause a general temperature rise. Hence we are able to relate the "heating process" of a toaster to a collection of energy-transfer nanobot operations (perhaps dedicated "heater" nanobots, or a side effect of inter-atomic bond-ripping ones, or both). We have the makings of a bridge to support an understanding of how the Cosmic Toaster works.

Similarly, the science of chemistry tells us that the essence of a toasted surface is a predominance of carbon atoms, and this can be produced from the hydrocarbons that constitute bread by removing hydrogen and oxygen atoms. There is a clear link between atomic-level activities of nanobots and our understanding of "the browning process" that produces a toasted surface.

Why is one holistic system (the Cosmic Toaster) open to the construction of a localised cognitive-level understanding of how it works, yet another holistic system (the MLP model of the brain) is not? In the first case we have a good grasp of the fundamental salient features, and in the second case we have little idea of what the salient features might be. And not unrelated to our confidence in the feature-salience, the sciences of the holistic level (atomic physics and molecular chemistry) appear to provide a sound basis for interpretation of the salient features at the holistic level. At the holistic level of brains (neuronal structure and function), the basis for interpretation in terms of what might be the salient features of intelligence, e.g., memory, is largely missing.

A more prosaic reason for the crucial difference between our two holistic systems is that toasting is undoubtedly simpler than intelligence. Additionally, I dreamed up the Cosmic Toaster's holism, so it's no surprise that it's understandable. Evolution delivered our intelligence by way of a brain. Consequently, no such similar implication of straightforward understandability is to be expected of holistic neural processes.

By adoption of a holistic computer-modelling technology, such as that of MLPs, does the structure-to-cognitive-level-behaviour chasm disappear? After all, we have used only evidence of the brain module's behaviour to construct a computer model. We have managed to bypass all aspects of the brain-module's internal structure.

By such means, the reverse engineer can construct a holistic computational simulation of the brain module scanned. Can this be a useful model in terms of our goal of constructing an Artificial Intelligence and by so doing, understand intelligence? Because the MLP model is guaranteed to behave just like the brain module, we can surely use it within our artificially intelligent system. And because we still have no cognitive-level understanding of how it does what it does, we'll just treat it as a "black box" that somehow does the job. Let's look at the big picture.

with a holistic model → jump across the abyss

scan-data model

| brain module | → | detailed brain-scan evidence | ⇠ - ⇢ | cognitive-level understanding of LIC1 process |

necessarily mirrors structure of brain module

?

MLP model

input-output training samples & MLP training algorithm

a programmed module that is "guaranteed" to reproduce the behaviour of the brain module, but no cognitive-level understanding → a "black-box"

duplicates input-output behaviour of brain module; no necessary structural correspondences to brain module

information processing:

localized symbolic at the cognitive level — X1, Y1, ...; squaring etc. according to Pythagoras' rule

information processing:

localized symbolic at neuron level or MLP-unit level

both distributed at cognitive level ➡ waves of information flow

➡ single threads of information flow

If all this is by and large acceptable, then what's the fallout? It tells the reverse engineer to forget efforts to analyse and reconfigure internal components of brain-module information processing, and instead accept it as a whole.

The brain-scanning technology is used to feed appropriate signals into each identified brain module and then the resultant output signals are recorded. In this way a collection of input-output signal pairs can be amassed for each brain module. This evidence of a brain module's behaviour can be used as a set of training samples for MLP technology to automatically generate a model that behaves just like the identified brain component. In this manner a collection of MLPs can be generated.

The intelligent software system that is built will then be an interconnected collection of MLPs each of which simulates a particular brain module as a whole. Each MLP will reproduce the behaviour of an identified cognitive-level process , such as LIC1. The construction of the complete Singularity system will be based on whatever is the fastest and most convenient hardware for replicating the input-output behaviour of the various brain modules scanned.

However, if this strategy is pursued exclusively (and we have as yet uncovered no other even vaguely plausible way to realise reverse engineering), the reverse engineer will end up with a collection of brain-module simulations, most of which will be "black boxes" — i.e., models that simply replicate the input-output behaviour of various brain modules with no understanding of either what these signals represent in terms of cognition, nor why they are so transformed.

Here is this modified reverse-engineering exercise:

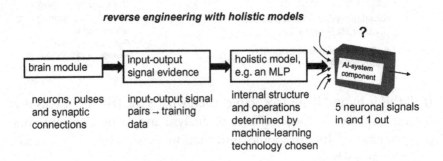

reverse engineering with holistic models

It may turn out that other information sources may occasionally give some cognitive-level handles on what a brain module is doing. For the most part, however, we cannot expect the luxury of a complete and certain semantic model such as the one we have benefited from in the reverse-engineering example exercises in this book.

Given a large collection of black-box and even some less-than-black-box simulations of various brain modules, how does the reverse engineer find out how to inter-connect them for the final achievement — an intelligent system?

It would be like trying to build a computer from a collection of electronic components about which the engineer knows no more than what signals each component accepts, and what output signals it generates. In addition, the engineer is trying to build the very first computer with little more to go on than vague hunches about how a computer might be configured and about what each component might contribute. To top it all off, there are no assurances that the pile of components he has to work with is either necessary or sufficient for the task.

A couple of decades may strike you as quite a long time to explore the possibilities for interconnecting components from within this pile of electronic black boxes (although, as yet, we do not have a single one of these components of the Singularity available, let alone the whole set). If so, then you've succumbed to the usual human propensity to vastly underestimate the escalation of possibilities when various options can be freely combined. We humans are terrible at estimating the magnitude of so-called combinatorial explosions, which leads inexorably to the unbelievable truth that **no computer will ever count 1, 2,... exponentially many** (Maxim 10).

The possibility of combining such black-box components is just one example of this ubiquitous false belief about very large numbers that we humans cannot shake off (except perhaps for some disciplined and rigorous mathematicians). This belief is reinforced by the amazing speed and storage capacities of modern computers; the reinforcement is powerful but illusory.

There are many apocryphal stories of primitive peoples whose counting skills stop at very few objects. They can count 1, 2, 3 objects but after that it's just "many". In our more demanding world where precise large amounts of, say, money, can be important, we've learned to go much further in our counting skills (some further than others, it's true).

By analogy with our con-specifics that live in less numerately demanding environments, our counting skill, yours and mine, can be summed up as: 1, 2, 3,... 100s, 1000s, millions,... exponentially many. But let's summarise this more sophisticated skill as: 1, 2, ... exponentially many.

So what's false about it? It is the belief that, with the power of modern computers, counting or processing any number of items will always be a possibility. If not today then tomorrow, given the fantastic regular improvements in computer power. It is easy to demonstrate, but still hard to believe, that this is false.

The importance of this human weakness arises in several different guises within our exploration of intelligence — e.g., the black-box system building that we've just outlined; the much earlier extraction of patterns of information processing from brain-scan evidence; and the attempt to attach a meaning to the output of LIC1 once we knew the meaning of its inputs. Combinatorial explosion of possibilities is a consequence of exponential growth, a phenomenon of the supporting technologies (such as computer power) so welcomed by Singularity advocates. So let's see exactly what we're up against.

An example I like is the tale of the Chinese emperor who wished to reward the valuable services of a local sage:

> "What gift would you like? Ask anything of me!" the Emperor declared.
> "Some rice," replied the canny old man.

"Some rice!" spluttered the Emperor, "Not gold, or jewels, or pearls?"
"No, just one grain of rice on the first square of your chessboard, two on the
second, and keep doubling the rice for each of the 64 squares. That much
rice will be quite sufficient," the sage calmly stated.
"Well, if that's all you really want," said the perplexed Emperor, "the rice
is yours."

The Emperor did not know just how much rice he had promised.
More than all the rice in China — much, much more.

Still in an aside but highly pertinent to the supposed meteoric rise of
the hardware technologies, we might note that the essence of such
combinatorial escalation is exponential increase. The doubling of
rice grains makes the total requested equal to $1+2+4+8+16$, etc.,
which is $1+2 + (2\times2) + (2\times2\times2) + (2\times2\times2\times2)$, etc. I trust that
this formulation of the sage's request is acceptable, and to help we
have the picture with number of rice grains on the first and last few
squares explicitly stated.

As illustrated, in mathematical notation the total of rice grains requested is: $2^0+2^1+2^2+2^3+2^4$, etc., up to $+2^{63}$. The small superscript numbers are known as "exponents", and as this series makes quite clear, it is the exponent that increases as we move from one chessboard square to the next. Hence, we say that the amount of rice promised grows exponentially as the chessboard is traversed. The last of the 64 squares will contribute 2^{63} grains of rice to the sage's reward (because the first exponent is 0, not 1). How much rice is that? Quite a lot you might guess, maybe a few sacks full, or given that the sage deserves his title, he might have sneaked a lorry load pass the perplexed Emperor?

Let's see: 2^{63} is about 10^{18}, and world rice stocks in 2009 were reported as 91.5 million tons[18], or 2×10^{11} lbs, and I think it is safe to assume that this will be more than the world production total whenever our mythical Emperor was dispensing his largesse.

So how much does the last chessboard square add to the sage's take-home gift of rice? It's about 10^{18} grains of rice. How much is that? It's $10\times10\times10\times...\times10\times10\times10\times10$, 18 tens multiplied together.

1 lb of good basmati rice (we assume that the Emperor was not a cheapskate) contains about 41,000 grains (according to my count). Thus the 2009 world production was 2×10^{11} multiplied by 41,000 grains of rice — that's a bit less than 10^{16} grains of rice. Now 10^{16} must be multiplied by 10^2 to give 10^{18}, and 10^2, $10\times10=100$. So the world rice stock for 2009 was 100 times short of the last square's contribution!

Rather than a few sacks full, the Emperor had promised his advisor the total world rice production (at new millennium levels) for more than 100 years! But this was only the last square's contribution, the other 63 squares will add the same amount (minus one grain[19], but let's not be too picky). Hence, the Emperor will need to corner the world rice market for at least 200 years in order to deliver on his promise!

Beware of exponential increases. They will appear as large, but innocently manageable numbers; they are not. Hence the significance of the false belief in the manageability of 1, 2,... exponentially many. Exponential growth may promise virtually limitless gains in time or resources, but they ensure that in a world where nothing is unlimited, whatever the resource being used, it will run out surprisingly quickly.

But back to the task at hand — maybe there are some clear, simple and very limited connections to be found between certain brain modules, but many, if not most, are likely to be buried within a dendritic tangle. Who knows, and it only takes one such incomprehensible tangle to derail the grand plan! Composing the final simulation from all the bits and pieces is surely going to be a challenge (to put it mildly), and yet again the Singularity proposal ignores it. The hope seems to be that the discovery of global underlying principles will do the trick. Well maybe they will, and maybe they won't. A strategy that rests on the expectation that this will all become clear, and fall on the positive side, takes hopeware up to a new level.

Holism accounts very nicely for our lack of success with developing a cognitive-level information-processing model of the small and simple process called LIC1. Although it might be a shame that holistic systems by their very nature will obstruct the development of a cognitive-level understanding of what's going on inside them, on such a small scale this absence of scientific insight is perhaps tolerable. The geometric competence, for example, of our simulated intelligence could be composed of many black-box modules that each deliver one of the processes that collectively are this competence.

Medical science, as well as you and I who are its beneficiaries, often accepts this level of understanding. The black-box "behaviour" of certain medicines is well-known and understood — if you take this pill for a headache, the headache disappears. In general, medical science will have a more or less (more for some medications and less for others) tenuous grasp on exactly how and why the medicine works.

The biggest worries surface when several such medications are combined. The unknown operational details may interact in surprising and negative ways[20]. Consider how this concern will escalate when dozens or even hundreds of incompletely-understood medications are administered simultaneously.

Of course, they never are, but the Singularity engineer will be faced with just this possibility (perhaps inevitability) when attempting to combine many similarly ill-understood holistic computational models in order to construct an intelligent system. What hope is there that he will be able to "manage" all the surprising side-effects that just keep popping up? None at all, is the answer.

Here's his problem:

engineering with black boxes

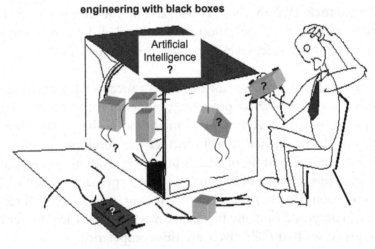

The Singularity talk that touches on modules tends to be of brain regions, not small brain modules. It is also explicitly stated several times that the eventual modelling will not be at the level of replicating individual neurons, but clusters of neurons.

It might then be objected that the reason our attempt to reverse engineer the small and autonomous brain module has repeatedly failed to live up to expectations, no matter what we assumed and what we did, was because we focussed on individual neurons. We were scanning and modelling at too low a level. We need to investigate the neural clusters that may well exhibit recognisable elements of intelligent processing (although LIC1, despite its small size, has a good claim to the label "intelligent processing").

Just as the close monitoring of the activities of single bees may fail to reveal the essential functionality of the hive, so the focus on individual neurons may be undermining attempts to understand intelligence. It is patterns of collective worker-bee behaviour, such as foraging for nectar, that are really what is wanted to inform an understanding that would lead to a simulation of the bee colony.

So, perhaps our mistake was the focus on single neurons, which is (despite well-known exceptions, e.g., in the optic tract) too low a level to find any cognitive-level principles. We need scan data for the activities of clusters of neurons, perhaps thousands. Although it is unclear (but by no means beyond the wit of anticipated nanobots) how to generate scan data for certain neuronal clusters without interference and noise from other adjacent and/or inter-connected ones, this must be a possibility — one whose solution is well underway, I expect. Electro-Encephalograms (EEGs) have long provided scan data for neurons en masse, and this technology is always being improved (probably exponentially).

Can this sort of scan data be a bridge from brain anatomy across to cognition? Perhaps in terms of holistic brain-region simulations?

We do now have a holistic MLP model that is guaranteed (as much as anything can guaranteed in empirical science) to behave exactly like the brain module, and can be built automatically from realistically obtainable brain-scan evidence. Is this black-box component, especially if built on a larger scale, going to contribute towards an understanding of intelligence?

Let's consider whether smallness itself might have scuppered the reverse engineering exercise. A larger brain module would presumably be capable of a variety of behaviours, i.e., it would compute other processes as well as LIC1. It might, for example, be the brain region that "knows" Pythagoras' rule as well as a number of other general geometric principles, and so could reason about a range of different geometry problems. Just being bigger than the LIC1 module without exhibiting more variety of behaviour could not possibly allow us to make headway. So multi-process behaviour is perhaps the missing feature of our earlier analysis.

The assumption of multi-processing begins to sit more comfortably with evidence that brains with geometric competence can indeed reason across problems and apply, say, Pythagoras' rule to a range of different geometric puzzles. It also avoids the conundrum posed by encapsulated modularity — the problem of maintaining belief in, and competence with, a general reasoning principle like Pythagoras' rule when it manifests as a variety of particular applications, each encapsulated in separate modules.

With the assumption of multi-processing brain regions as the major components of reverse engineering, are we free in one bound? No, the leap is one of the frying-pan-to-fire variety, because now we are really in trouble with our reverse-engineering project. Why?

To begin with, the scan-data evidence will relate to a variety of different processes making it even more difficult to interpret, i.e., which piece of evidence accounts for which component process? In addition, if our brain region is a holistic implementation of the various processes, then there is no known way to tease apart the individual processes. And from a purely practical standpoint, MLP-type learning technologies are strictly limited to single-process learning (as we shall see in a later chapter).

But if we are dealing with brain-region black boxes (assuming that we can devise a learning technology to accommodate multi-process modules), the interconnection possibilities just escalate further.

Maybe the reverse engineer will not be faced with a huge pile of neuron-level scan data (various input-output pairs) when dealing with a multi-processing brain region. Perhaps some sort of region-summary scan data can be extracted and presented for analysis?

EEGs have long been available to provide such "bulk" scan data. With scanning technology improving by leaps and bounds it is clear that refinements of EEG-type scanning will be able to give the reverse engineer focussed patterns of activity about what's going on within some identified brain region. But how is this going to make the task easier?

The brain region must be a multi-processing one because that's the whole point of switching from small, single-process brain modules to regions that account for more global characteristics of intelligence. Holism or no holism (if we're lucky), how can our region-averaged data hold more promise of interpretation in terms of a mix of unknown processes than our neuron-level data did for interpretation in terms of one simple (and known) process? It beats me as to what advantage brain-region data derived from a multi-processing module offers over our initial "single-module, single-process" exercise.

Whatever the appeal of holism as a characteristic of a well-understood computational technology, it soon switches to horror when a cognitive-level understanding is required. This is because distributed, holistic information processing is fundamentally at odds with the localised, compositional understanding desired — although, as the Cosmic Toaster demonstrates, holism is not necessarily unfathomable. Maybe MLP holism is just the wrong type, and we need to devise technologies that will generate more understandable models — perhaps Cosmic Toaster-like holism? Or perhaps we just need to pin down the salient features of intelligence. Or perhaps both?

Holism, like so many of life's intriguing temptations, can be attractive in small doses but it can all too quickly become too much of what initially seemed like a good thing.

The reverse engineer really does need the *underlying principles* of the information processing of the brain, so black-box modules are not good news. But the only general underlying principle we've found is the squashed-summation and weight-product transfer as captured in our earlier syntactic model. The move to consider bigger, more realistic, brain regions far from promising, or even hinting at, a way around our earlier impasse threatens to bring the reverse engineer nothing but bigger troubles.

Maybe we did not look hard enough, or intelligently enough? Where are we going to find intelligent guidance given that our goal is to understand intelligence and simulate it? This hint of circularity is manageable because the world of traditional AI has been developing general principles and specific models of many aspects of intelligent behaviour for more than half a century — mostly quite independent of brain-structure considerations. These models, which have overwhelmingly derived from analyses of cognitive behaviour, rather than observation of brains, may provide elements of guidance for the reverse engineer. They may provide insights into the salient features of intelligence, or even provide working models that will become integrated into the grand plan, the ultimate simulation.

We need to see what's out there, and how it might be useful.

Endnotes

[1] The quotation is from F. E. Sutcliffe's translation of Descartes' *Discourse on Method and Other Writings* (Penguin, 1963), p. 164.

[2] Jerry Fodor's *The Modularity of Mind* (MIT Press, 1983), p. 127.

[3] Such a claim is made about a mysterious property of "memory" in water that "explains" the medical benefits of pure (but previously treated) water within the murky world of holistic medicines. This is not to claim that science is aware of all valid forces and phenomena in nature. This "memory" phenomenon might exist, but so might any other of an infinite number of equally valid "explanations", all utterly devoid of any objective support. So why put faith in this particular guess? See Wikipedia (accessed 31/1/2012). "Water memory is the claimed ability of water to

retain a 'memory' of substances previously dissolved in it to arbitrary dilution. No
scientific evidence supports this claim. Shaking the water at each stage of a serial dilution is claimed to be necessary for an effect to occur. The concept was proposed by
Jacques Benveniste to explain the purported therapeutic powers of homeopathic
remedies, which are prepared by diluting solutions to such a high degree that not
even a single molecule of the original substance remains in most final preparations.
Benveniste sought to prove this basic tenet of homeopathy by conducting an experiment to be published "independently of homeopathic interests" in a major journal.
While some studies, including Benveniste's, have reported such an effect, double-
blind replications of the experiments involved have failed to reproduce the result. The
concept is not consistent with accepted scientific laws and is not accepted by the
scientific community. Liquid water does not maintain ordered networks of molecules
for longer times than a small fraction of a nanosecond."

[4] Another possibility to be pursued is that the hydrogen and oxygen atoms removed
from the slice-surface molecules are recombined in an exothermic process to give
water molecules and some heat. This heat is then transferred back to the originating
slice surface.

[5] Change any single character of the cognitive-level model of LICl to any other
character and this model will cease to behave correctly. By way of contrast, change
any link weight of the network model and it will almost certainly have no effect on
the correctness of its behaviour.

[6] All life forms are composed of a variety of different cells. Humans, for example, are
composed of skin cells, bone cells, blood cells and brain cells, some of which are
neurons, to name but a few. Most of the cells that constitute you and me regularly
die and are replaced by new versions. Human skin, for example, is completely
replaced regularly, perhaps every month or so. A major exception to turnover of bodily cells is our brain neurons. We are born with, or soon develop, our full comple-
ment, and for the next three-score years and ten they die regularly without
replacement. So there is a good argument to the effect that the persistence of our
cognitive capacities over decades and decades can only be explained by some sort of
holistic knowledge representation in our brains. However, just as with so many other
aspects of the human brain, the situation is not this clear cut: for example, in recent
years evidence has emerged for a degree of brain-cell regeneration in some areas of
the brain.

[7] I'm hedging here because at this atomic level of matter the Earth scientists might
well run into the fundamental indeterminacies of quantum phenomena or even that
of Heisenberg's "uncertainty principle". The visiting aliens, who constructed the
Cosmic Toaster, are clearly ahead of us with regard to managing the unpredictable,
but in ways that I cannot even begin to speculate about.

[8] This is not to claim that neuroscientists are totally ignorant of relationships between
brain regions and intelligent behaviours; it is merely to say that pretty much all such

relationships are known only in a very general sense. So, for example, the neuroscientist Tali Sharot, in her intriguing and informative book *The Optimism Bias* (Robinson, 2012), repeatedly informs us of brain-mind interrelationships but they are all typically of the form: "The frontal lobes [of the brain] are critical for functions... such as language" (p. 53), or that "the caudate nucleus, a cluster of nerve cells deep in the brain,... has been shown to process rewards and signal the expectation of them" (p. 136). How the frontal lobe neurons perform any language-critical functions, and how the caudate nerve cells process rewards in a way that might contribute to, for example, our undoubted expectation when "we believe we are about to be given a juicy steak or a hundred-dollar bill" is not yet known.

[9] *Kludge: the Haphazard Construction of the Human Mind* by Gary Marcus (Houghton Mifflin, 2008).

[10] Richard Dawkins' book *The Blind Watchmaker* (Longman, 1986, and then Penguin, 1988). It is necessary to surmise here because Dawkins steers well clear of what is to be expected in terms of cognitive architecture as a result of the evolution of the human brain.

[11] We'll visit this technology properly in a later chapter. For the moment we might just note that the MLP architecture and training algorithm (known as the BP algorithm, for Back-Propagation of error) are guaranteed to converge on an optimum solution, and are, in principle, capable of learning any single function. It is these guarantees that I allude to by calling this technology "well-founded" in contrast to the dozens of other machine-learning technologies (including neural-net ones) that can support impressive demonstrations but are equally (or more) likely to fail dismally because they are not well understood. These technologies are hit-and-miss, try-it-and-see; if it fails, try something else — not a sound basis for science.

[12] Two typical squash functions are illustrated on the next page; they are S curves — a distinctly S-like solid line and a more angular dashed "curve". The "hard" thresholds that we have hypothesised can be considered as squash functions in which the slope of the central section is vertical. Thus the dashed squash function illustrated is close to a hard threshold (or step) function centred on an input value of zero. If the central dashed portion were vertical on zero then the function would be: if input signal greater than 0 *then* output signal is 1, *otherwise* output signal is nothing.

Any input value, which will give a position on the horizontal axis (the value 0.65 is shown as an input to this function), is transformed into a value on the vertical axis according to the detailed shape of the S curve used (the output shown is 0.75 for one curve and 0.9 for the other). Notice that while the input axis accepts any value, the output axis is limited by the S curves to a maximum of 1.0 and a minimum of 0.0. So any input signal is "squashed" to an output between 1 and 0. Notice also that the shape of the curve, primarily its steepness, determines how closely to 1 or 0 each input is squashed.

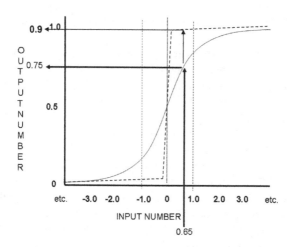

With a squash function, we can do away with the threshold values but we get them back again in the form of values for the slope of this S curve. It may, for example, be very steep as illustrated by the dashed example or it may be less steep (as in the more S-like curve illustrated). But such details will be glossed over for now, and revisited in a later chapter where we examine the "new" technologies. Suffice it to say that within an MLP unit there is "squash"-type information processing, and, whatever the internal sum of signals, it generates an output signal with a value between 1 and 0.

[13] It would be sensible, and generally correct, to assume that the brain-module signals recorded will more easily reveal their inherent patterns when their accuracy (some of which is likely to be spurious) is reduced by rounding the values. By so doing, however, the neuron output signals that are typically close to 1 or 0 will be rounded to 1 or 0, respectively. Thus, 0.9957 becomes 1.0, and 0.0012 becomes 0.0. Such rounding may be useful if the biological neurons are best interpreted in terms of "pulse" or "no pulse" output signals, but if not (and for an MLP certainly not), the rounding unhelpfully distorts the data. As with so much of the reverse engineering proposal, circular reasoning intrudes; knowing quite a lot about how to best interpret the scan data is necessary for correctly interpreting the data.

[14] This automatic weight-adjustment algorithm, whose discovery in about 1980 launched MLP technology, is called the "Back-propagation of error" algorithm, hence, the Backprop, or BP, algorithm.

[15] I use the term "learning" but given that, in the MLP context, a full set of correct outputs must be known, it is perhaps more accurately a process of "training" to reproduce these known output signals.

[16] There are a number of differently shaped S-curves that can be used, and any given type of S-curve can be varied by adjusting the slope of the central section. Two possibilities are illustrated above: a steep sloped, angular curve, and a less steep, more

175

S-shaped curve. So, there are many different squash functions, and one must be chosen by the experimenter.

[17] The link weights in this model have been dreamed up for the purposes of illustration. They are not values that have been learned. So, although this illustration looks like a trained MLP, it will not actually compute LIC1, but we have constructed many similar MLPs that do compute LIC1.

[18] The tonnage of world rice stocks, as well as many other fascinating rice facts, is accessible at www.rice-trade.com (accessed 6-6-2011).

[19] For those readers who are getting to like this mathematics stuff — the other 63 squares will contribute $2^0 + 2^1 + \ldots + 2^{61} + 2^{62}$ grains of rice to the sage's take-home total, and this is $2^{63}-1$. Why? Because there is a general formula for summing the terms in such series. When it is this type of doubling series, any term in the series, say 2^n, is just one short of being equal to all the previous terms from 2^0 up to $2^{(n-1)}$ added together. Try it: 2^4 is $2 \times 2 \times 2 \times 2$ which is 16, and the previous terms are $2^0 + 2^1 + 2^2 + 2^3$ which is $1+2+4+8$ and that sum equals 15. This single test proves nothing, but you can assure yourself that my general assertion might be true by trying as many more tests as you wish. Ultimately, of course, if you want solid proof then you have to look into the formula for summing such "geometric series".

[20] The thalidomide tragedy of the 1950s in the UK was the result of just such an unanticipated interaction — thalidomide is a sedative drug that was used to treat morning sickness and to aid sleep. In the absence of a full understanding of how it worked, its ability to cause birth defects went unnoticed until too late. It cured the pregnant woman's nausea but it also had a disastrous effect on the developing foetus.

ଓଃ Chapter 7 ଡ଼

Hoping for a Knee up Soon

"We are now approaching the knee of the curve (the period of rapid exponential growth) in the accelerating pace of understanding the human brain."

Ray Kurzweil, 2005[1]

"Within a generation the problem of creating 'artificial intelligence' will be substantially solved."

Marvin Minsky, 1967[2]

"We are no closer to a computer that thinks like a person than we were fifty years ago."

Marc Andreesson, ca. 2009[3]

AI is not dead, definitely not. Over the last half a century or more, the field has moved along in fits and starts — one might say it has moved along steadily, although others might prefer "sluggishly" as more apt. No one without a vested interest, or devoid of technical competence, is likely to be a great deal more positive.

Despite our concentration on the human brain, the vast bulk of AI research projects have ignored this organ. Why? Well, if it's just a "machine" developed by evolution on which is installed the intelligence program, and it is moreover a sub-optimal machine for this task, then it makes sense to ignore it. If there is at least a modicum of truth in the foregoing, then the "program" that is human intelligence will likely have had to make concessions to its processor, the brain, and so be sub-optimal as well.

You forget things, and I do too. On the face of it, this characteristic of our intelligence is an example of sub-optimality. Why does it happen? We do not know, but it is plausibly due to the manner in which our brains store and retrieve information, i.e., a weakness of the underlying machine.

Ever since the problem-solving abilities of the first shrimp were centred on a small neural tangle in their heads, the basic component of the organic thinking machine was set — a network of neurons. In the passing millennia, life forms evolved to become more sophisticated thinkers, and neuron-net organisation was repeatedly expanded and revamped, but the basic building block was not.

In contrast to organic brain technology, computer hardware technology has leaped forward several times within the span of a human lifetime. It therefore provides us with a concrete example of the repercussions of creative re-design in contrast to making do with the first solution.

In 1964 when I entered University College, London, to study for a Degree in Chemistry, the X-ray crystallographers led by Professor Dame Kathleen Lonsdale were a prestigious research team. These researchers used a Pegasus computer to perform the intensive calculations needed to help them interpret the X-ray diffraction images of their crystals, and so shed light on their molecular structure.

I still clearly remember Professor Lonsdale impressing us freshers with stories of automatic computation. In particular, she explained that a big problem was keeping the Pegasus computer running for long enough to complete a calculation. The crux of the problem was not that the calculations took an inordinately long time, but that the computer frequently broke down. This happened because the basic on-off electronic switching component was the vacuum valve — a large, hot and unreliable device that looked a bit like a light bulb.

Every time one such valve ceased working properly, the Pegasus computer ceased working altogether. The faulty valve had to be located, and replaced before the computer and the calculations could be restarted, probably from the beginning (I've forgotten some details).

The photo on the right shows a couple of the Pegasus computer's vacuum tubes.

Suppose that after half a century of progress in computer design and development, these impressive glass valves were still used as the basic switching device. They would probably be smaller and more reliable, and organised into networks containing multiple alternative processing routes to provide a redundancy that would help counter their fundamental unreliability.

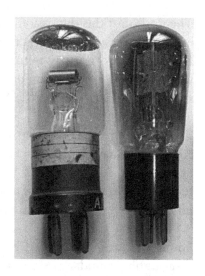

But this single point of failure to improve on the basic technology of the fundamental binary (i.e., on-off) switch would mean that many of the conveniences we now enjoy, such as Blackberries and smart phones, would not have emerged. The neocortex, that late addition to the brain, may well have a sophisticated, hierarchical layer and columnar structure that takes care of much of our intelligent behaviour[4]. But it is no more than an expansion and reorganisation using the shrimp's neuron. So, is intelligent thinking a process that is working as best as it can on a machine that is built from outmoded and antiquated components?

Is your brain like a desktop computer composed of a large cabinet full of glowing glass valves which requires a hard-working refrigerated cooling unit? If so, it may be that the neuron-based information storage facilities or information-retrieval processes, or both, necessitate sub-optimal processes in our intelligence program. For example, your brain is a machine that, despite the engineered redundancy, loses previously acquired information on a regular basis. Modern computers, by way of contrast, do not forget anything (leaving aside the occasional catastrophic disk crash).

179

The upshot of this digression is that most scientists who are attempting to understand intelligence and build computer models of various aspects of it do not pay much attention to the human brain. If you want to build a new Formula 1 car, you do not waste time understanding the intricacies of the Model T Ford.

Even freed of the constraints observable in brain-based intelligence, AI projects have not been runaway successes. It is true, to some extent, that AI researchers have been repeatedly robbed of their recognition when mainstream software engineering adopts the latest innovation while at the same time denying its AI provenance.

Some balance is introduced by the adoption of the "AI" label for a thoroughly mundane software product in order to secure the associated kudos — a step up in comparison to competitors, and so a correspondingly better market price.

Jaron Lanier, a pioneer of virtual reality, sums the general mindset up for me by saying, "Whenever a computer is imagined to be intelligent, what is really happening is that humans have abandoned aspects of the subject at hand in order to remove from consideration whatever the computer is blind to."[5]

A tongue-in-cheek observation that I've heard attributed to the AI pioneer John McCarthy concerns the intelligence of the vacuum flask — not only does it keep a hot liquid hot, but it knows when its contents are cold and then it keeps them cold.

No small part of the squabbles over how well AI has fared has to do with definition[6]: what constitutes an AI system? The field itself has sometimes tried to draw a line between strong and weak AI — the former, a computer system that simulates perhaps all of human intelligence, and the latter, systems that demonstrate some aspects of intelligence such as learning, or natural-language communication. Needless to say, strong AI has always been seen as a distant prospect. One that disconcertingly has faded increasingly further into the wide blue yonder as the full magnitude of the deep complexities of human intelligence has become apparent.

Cheerleaders for the Singularity — strong AI and more — have been wildly upbeat. They see traditional AI systems as romping along towards the knee of the curve for imminent exponential lift-off. How do they justify this glowing vision? This is achieved partly by redefining AI, and partly by use of that tried and trusted device of optimistically extrapolating over the horizon from current inadequacies — in a word, hopeware. Ever since Alan Turing, who was introduced in Chapter 1, and the other pioneers of modern AI began to gaze optimistically beyond their initial insights and into the future of AI, hopeware has been the vehicle of choice for the major breakthroughs. Its popularity has never flagged, and so it merits its own chapter later in this book.

The definitional ploy, as run by Kurzweil, introduces a new category: "narrow" AI which is "creating systems that can perform particular functions that used to require the application of human intelligence" (p. 92). Under such a rubric, AI, in its "narrow" form, is doing well and has been romping along ever since that meeting of minds between the first programmer, the Countess of Lovelace,[7] and the computer pioneer, Charles Babbage. Although undocumented in the historical record, it's a good bet that these two began to plan the details of "programming" Babbage's brass-cogs, wheels and spindles computer to add together two numbers. Number addition is, of course, one of many "particular functions that used to require human intelligence", and still does, of course, when an adding machine is not available.

With a definition that makes AI, in its narrow manifestation, virtually conterminous with "programming", it is of course doing famously. (I've stopped short of claiming equivalence here only because I feel that there must be software that falls outside the narrow-AI definition, but I'm struggling to come up with software system that is not performing any functions "that used to require human intelligence".)

Anyway, under this loose licence it would seem perfectly legitimate to state that "we already have an effective toolkit for narrow AI" (p. 293), although the software engineer fighting his losing battle with the emergent chaos of a large IT system might want to hold back a bit on even this level of optimism[8].

A history of scaling down was how I characterised AI research in 1991[9]. With decades of AI research having flowed out of both academia and entrepreneurial labs, it is still true that, despite definite elements of progress, no exponential blossoming of AI progress is evident. The big picture is more of little bubbles that promise much soon, but burst or just fail to grow properly. The knee of this particular technological development curve may be just over the horizon... and it looks likely to stay right there.

To take a relevant case, the Turing Test — Loebner Prize competitions have been running regularly for several decades, but have we seen the new systems getting progressively closer to passing the Turing Test? I think not. I attended the Loebner Prize competition held at the University of Exeter in the UK in March 2011. As one of the interrogators confirmed, it was immediately obvious which system was the computer. And later, at a question-and-answer session with Hugh Loebner I asked if he saw improvement in the computer systems over the years. He said, "Yes." But as was his style he offered no justification and immediately veered off into unrelated matters that he wished to air[10].

Way back in 1972, Hubert Dreyfus[11], a philosopher by trade, delighted in detailing the over-extravagant claims of AI researchers, and how they get further inflated as they pass through subsequent commentators, until the sky's the limit once the popular media latch on.

Among his many illustrations of how "the intellectual fog" is produced in AI, he starts with the prediction in 1957 by Herbert Simon[12] (a leading light in AI research for decades thereafter) "that within ten years a digital computer will be the world's chess champion". In the following year, Simon and two colleagues announced that their "not yet fully debugged" chess program, whose behaviour one "cannot say very much about" was, nevertheless, "good" at game openings. Another year later and Norbert Wiener, an eminent cyberneticist, escalated the claim that this chess program was "good in the opening", adding that "chess-playing machines as of now will counter the moves of a master game". Another participant at the same meeting reinforced the escalation with the assertion that "machines are already capable of a good game." In fact, Dreyfus records, the "program played poor but legal chess, and in its last official bout (October 1960) was beaten in 35 moves by a ten-year-old novice" (pages xxix-xxxi). Yet the hype continues unabated.

Ever since IBM's Deep Blue chess-playing computer beat World Champion Garry Kasparov in a chess match in 1997, the end of human supremacy in this (unusually well-defined) field of intelligent human endeavour has been predicted — soon. Kurzweil — ever ready to exercise his clairvoyant abilities — stated in 2005 that chess-playing "programs running on ordinary personal computers will routinely defeat all humans later in this decade" (p. 278). True to form "this decade" has come to a close without the predicted breakthrough in pattern-matching to power the anticipated burst of computer prowess in chess playing[13].

No open-minded scientist is foolish enough to assert that this will never happen. The trick, which the Singularity proponents have failed to master, is to cast one's predictions into the more distant future. It is the height of folly to predict the near future. A decent distance beyond a human life-span, at least a working-life-span — this is the closest to home that hopeware can be sensibly sited.

Deep down, though, I suspect that the Singularity advocates know that this particular curve of technology shows absolutely no signs of lifting off in the foreseeable future. Hence they throw (most of) their eggs in the reverse engineering basket. Why not all? Why bother with even lip service to mainstream AI?

My suspicion is that the sought-after pattern-recognition principles that (apparently) underlie human intelligence are not going to leap out of the brain-scan evidence, and the Singularity believers know this too. But if these crucial principles of information processing are hinted at, if not revealed, by traditional AI projects, then perhaps they can be confirmed or finalised in the models abstracted from the brain-scan evidence.

In summary, traditional AI may provide some guidance and direction to the process of brain-scan evidence interpretation at the cognitive level. A sort of Moses to lead the reverse engineer through the wilderness of exponentially many possibilities for patterns in the data, and a similar 40-year journey to the promised patterns would be truly remarkable. An expectation of less than two decades is just ridiculous.

Curiously, the exponential-growth mania appears to be all embracing. The sluggish progress of traditional AI must be presented as yet another technology about to take off. Tellingly there are no impressive graphs of AI projects rocketing the technology into the stratosphere.

Having made much of the definitional dodges used to present whatever picture of AI is desired, I feel the weight of an obligation to participate in the game. So, rushing in where others, less foolhardy, might choose to tip-toe around, I'll give you my take on this issue.

Intelligence, it should be painfully clear, is a slippery notion. Pinning it down is like nailing jelly to a tree. As far as we know, humans have an absolute monopoly of intelligence in the here and now as opposed to "soon". AI enthusiasts, and I'm right behind them here, believe that the wetware of the human brain just happens to be the only known substrate for intelligence. They believe that there is no necessary relationship here. Indeed, the biological brain, as we have seen, may well be a less than optimal substrate. So it should be possible to recreate intelligence on other substrates, such as a computer. It may also be possible to improve on the human version — perhaps faster or less forgetful.

There is much, and probably endless, argument about most of my assertions in the preceding paragraph. Taking a broad view of human intelligence, for example, excursions into human emotions such as love, and feelings such as pain, can suggest that intelligence, in its human manifestation (the only one we know), is inextricably bound to having a body, and maybe a "blood and guts" brain.[14]

The underlying hypothesis of the AI researcher, which you may recall from the opening chapter, is called the Computational Theory of Mind (CTM). I have crudely summarised it as the belief that intelligence can be manifest as a computer program, a software system, running on a computer.

Just in case your particular installation of the intelligence program has already dropped this earlier recollection, here it is again in pictorial form.

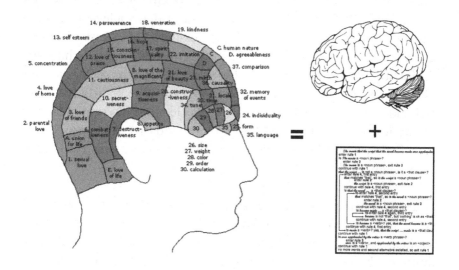

And now, when equations are part and parcel of your everyday reading, here is the more succinct formulation of the CTM.

MIND = BRAIN + PROGRAM

As also stated earlier, this view is open to all sorts of objections. If, however, we take a sufficiently limited view of intelligence as the integrated set of behaviours such as visual and auditory perception, natural-language understanding manifest via hearing and speaking competences, reasoning abilities, etc., the CTM strikes me as a viable contender, and I'm going along with it[15].

The Singularity project (tacitly at least also going along with the CTM) expects much more from its simulation but, as stated earlier, I do not. In this book we are working with a pretty limited view of intelligence, but one that is quite complex enough to demolish the grand-plan strategies. It cuts them off at their roots well before we need consider the possibility of the bells and whistles that will constitute "full" human intelligence, let alone a super-intelligence.

Before we move on, mention of the CTM opens the door for another illuminating peek at the "structure alone cannot deliver meaning" basis for my primary objection to the possibility of reverse engineering intelligence from brain scanning. The reader may well think my distinction between structure and meaning is arbitrary, petty and not one that dwellers in the real world ever need to be confronted with. It is nothing more than ivory-tower academics bickering, a purely domestic dispute — but then (as Fodor reminds us) so was the Trojan War. Let's see this distinction as it might arise close to many homes in your neighbourhood.

Your old smart phone runs a useful but costly App, one that you'd very much like to transfer to your new, improved smart phone-plus. To help you make this transfer your old phone has been thoroughly infiltrated by slick and informative nanobots. They ramble over all the circuits, processors and any other hardware components that your smart phone contains.

These miniature monitors report all the electronics they find as well as the full details of the signals that are whizzing around. That's a lot of data, a lot of information, but it's only evidence pertaining to the *structure* of your old smart phone, and the details of how the App causes this particular structure to work internally. It can only lead to a model of what your old smart phone does (whizzing signals around), and how it does it (the way the electronic components transform the signals they receive).

What chance is there that you could, by studying this nanobot evidence, reverse engineer the desired App even when you know exactly which one you're after — i.e., you know the general meaning of what your smart phone is doing when this App is running?[16] Lifting the burden off you personally, what chance is there that the program that is the sought-after App could be deduced and re-constructed from the scan evidence? No chance at all. Why not?

The nanobot evidence — the signals whizzing around, the values being stored, etc. — cannot be mapped onto elements of the original program, such as variable names and keywords, nor does it offer any way to apprehend program structure. Why not? Because these program details are not in your smart phone. They existed in the original program only as a necessary aid to human construction and comprehension of the App constructed as a general program that can be installed on a variety specific smart phones. For your particular smart phone running the App they are irrelevant, and so they do not exist. It is this "lost" description, a transferable one independent of the hardware, that constitutes a cognitive-level meaning of the desired App.[17]

Here's the story in pictorial form:

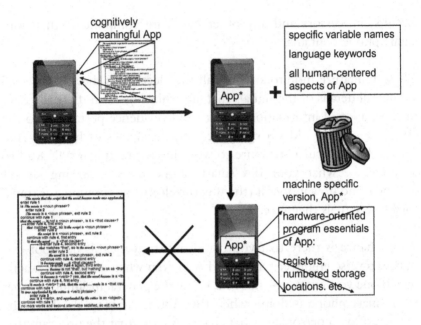

The essence of this chasm between structure and cognitive-level meaning is found in the wealth of human-centred information required for a human to design and construct the original program that is the general App. Once this information, which the smart phone does not need, is discarded, and the general App program is restructured to be efficient on a particular smart phone, it cannot be recovered. It's gone. So the original version of the App to install on the new hardware of your smart-phone-plus cannot be extracted from your old smart phone.

It's easy to throw information away, and it's then impossible to retrieve it. Here's the picture:

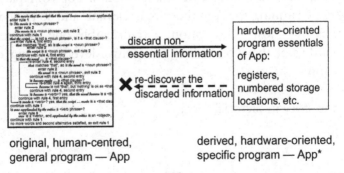

original, human-centred,   derived, hardware-oriented,
general program — App   specific program — App*

If you can accept the foregoing then substitute "human brain" for your old smart phone and "intelligent behaviours" for the costly App, and the gulf is made manifest. But back to the quest for the App in the brain that is intelligence, or would be if we could just extract it and install it on the latest electronics, or bio-computer, or whatever seems to work best.

So, what's my "definition" of intelligence? Learning, or self-organisation, is the key characteristic of intelligence, if I push further for simplicity and limit the choice to just one[18]. Why? Because non-adaptive intelligence is an oxymoron. A non-seeing intelligence is not; neither is a non-speaking intelligence. A non-thinking intelligence would be a worry, but thinking is close to a synonym for intelligence, whereas self-adaptation is not. (But notice the clash between learning as the key characteristic and our earlier requirement of no-learning as an essential aid to brain-scan evidence interpretation.)

A natural-language system, for example, must be capable of adapting its subsequent utterances according to information it gets in its earlier ones — even sentence to sentence. As both you and I are well aware, a totally fixed and inflexible natural-language system may have some limited success in hiding its lack of significant self-adaptivity by incorporating elements of (pseudo-) randomness (to, say, vary responses to a repeated question). But this and other such tricks will not last in the context of an extended dialogue. A non-adaptive natural-language system will never pass the Turing Test, except possibly as a one-off freakish event.

I do believe that a "learning" capability is at the heart of intelligence, and so it does begin to work quite well as the acid test of an AI system. It must be acknowledged, however, that the term "learning" covers a wealth of self-organising options, and so still offers wriggle room for dubious-AI to slide in through the loopholes. Any software system that makes changes to its organisation, even pre-programmed ones, or trivial, externally-triggered ones (such as adding new data

189

items to a list), could claim the "learning" accolade, and so would have to be admitted to the AI corpus.

It must also be acknowledged that if a "learning" capability is to be the acid test of AI then many worthy projects (most of those described in AI textbooks) suddenly become excluded from the oeuvre. In many cases, this retreat of the frontiers of what can be labelled "AI" is to be welcomed.

On the other hand, there is much good AI research to be done on the core capabilities of various intelligent behaviours well before the extra complexity of self-organisation is contemplated. However, the fact that learning aspects are often viewed as an add-on after the central problem has been cracked may be part of the overall problem with many would-be AI projects. Learning may well be *integral* to all intelligent behaviours, rather than a later add-on.

The Singularity believers are keen on self-organisation; indeed they welcome its advent. They see a self-organising system as one that promises a way around the awkward failures to understand various elements of human cognition — the recalcitrant modules will re-organise themselves appropriately and so circumvent their built-in inadequacies. "Fat chance" is the measured response of the software engineer who has grappled with machine-learning algorithms and the consequent software systems that keep changing themselves. It is time to tackle the pros and cons of self-organising systems — systems that exhibit the bare minimum to justify the epithet "intelligent."

Endnotes

[1] *The Singularity is Near* (Duckworth, 2005) by Ray Kurzweil, p. 154.
[2] Marvin Minsky, a leading AI researcher, stated this on p. 2 of his book *Computation: Finite and Infinite Machines* (Prentice Hall, 1967).
[3] Marc Andreesson, Internet entrepreneur, as quoted in *Googled* (Penguin, 2009) by Ken Auletta, p. 327.

[4] For example, Jeff Hawkins (with Sandra Blakeslee) in their book *On Intelligence* (Holt, NY: 2005) base a "new theory of intelligence" on the organisation of the neocortex.

[5] Stated on p. 35 of his book *You Are Not a Gadget* (Allen Lane, 2010) in which Jaron Lanier muses about many aspects of modern computer technology. In particular, he is sees much wrong with the current enthusiasm for "computing clouds" where large numbers in the algorithms supposedly remove the risks associated with creativity via statistics. Wikipedia, for example, seeks to erase individual points of view entirely and aspires to a superhuman validity via "the hive mind". For Lanier, the fragments of human effort that have flooded the Internet generate a flat mush of knowledge that cannot substitute for, let alone supercede, the creativity of an adventurous individual imagination that is distinct from the crowd. But "the hive mind" as a mass of neurons may well deliver basic cleverness which is all we seek to understand — see later fantasies about Watson playing *Jeopardy*.

[6] Many jokey definitions have done the rounds. For example: if it's written in LISP, it's AI. The LISP language was defined by John McCarthy ca. 1960 as a language well-suited to AI because it facilitated programming with associations between items that might be learned facts or processes. For a decade or so (roughly the 80s) LISP machines (specifically engineered to be efficient computers for LISP programs) were top of the shopping list for every serious AI laboratory. Nowadays, museums of technology are where you may find a LISP machine, and neural computing may have moved into the slot vacated. Neuromorphics, which refers to a new class of "brainlike computers." John Markoff in the New York Times on Dec. 29, 2013... "2014 could be the year that engineers and artificial intelligence researchers start to create computers that think and act like humans because their 'brains' have been engineered and designed to resemble those of humans." And pigs might fly.

[7] At this early date (ca. 1830) the daughter of Lord Byron, Ada Augusta, was not yet Countess of Lovelace, and Charles Babbage's "computer" was unfinished, but work on it was "well underway" (although never completed). B.V. Bowden's book *Faster than Thought* (Pitman: London, 1953) describes the scene when these two pioneers met as originally told by Mrs. De Morgan, wife the famous mathematician.

[8] In *The Seductive Computer: why IT systems always fail* (Springer: London, 2011) I invested most of the 322 pages explaining the ineffectiveness of the software engineer's toolkit for managing IT-system development and subsequent usage.

[9] In my *New Guide to Artificial Intelligence*, (Ablex: NY, 1991, now Intellect: Bristol) "A History of Scaling Down" is the heading of an introductory section that begins on p. 8.

[10] In mitigation, it has been suggested to me (by a Loebner Prize winner) that after winning this competition the tendency is not to re-enter again in later years (with presumably a better system). However, if there were any chance that an improved system might actually pass the Turing Test (rather than just secure the Loebner Prize once more — the least worst entrant) then surely the research team would re-enter

their project in order to capitalise on the undoubted celebrity (and consequent funding) that the supreme achievement would garner.

[11] Hubert Dreyfus published *What Computers Can't Do: a Critique of Artificial
Reason* in 1972 (Harper & Row). It was a swingeing attack on all that the AI
researchers held most dear, and from an outsider (a philosopher) almost within their
ranks (The University of California at Berkeley adjacent to the epicentre of AI activity
around Stanford University and Menlo Park, although MIT, across the continent,
spoils my neat picture). Although he swung widely, and was rightly chastised for
sometimes missing the mark, his documentation of the wide disparity between the
published claims and the reality of AI projects was hard to dismiss. He soon became
a resident critic within the AI establishment.

[12] According to Pamela McCorduck in her book *Machines Who Think* (Freeman,
1979), pages 188–189, Allen Newell was an equal partner behind Simon's pronouncements. This famous predictive episode in the annals of AI is dealt with fully in
the later "Hopeware" chapter.

[13] Brute force was the "secret" of Deep Blue's success, and Kurzweil was anticipating
something more edifying to power the AI breakthrough; we will revisit this particular
AI ploy in the closing chapters when we consider the latest IBM PR coup in AI —
"Watson", the *Jeopardy*-playing system of 2011.

[14] A persuasive argument along this line, which we introduced in Chapter 1, is that
sweetness and its consequent pleasure are *not* in the sugar molecule; these feelings are
generated by our brains. All such pleasant, and unpleasant, feelings have evolved in
our particular brain architecture. They have done so (according to the "emergent
mind" hypothesis, see Endnote referring to Victor Johnston in Chapter 1) in order
to distinguish between, and react appropriately to, a diverse range of potential threats
or benefits to our survival. If true, these feelings, which even basic intelligence perhaps cannot ignore, are then determined by some aspects of brain structure and
function.

[15] Even if we accept the "emergent mind" hypothesis, which requires that some
aspects of brain architecture are integral to our intelligence, this does not mean that
all of the brain is similarly vital. It would just mean that the necessary aspects are
identified and simulated in the underlying computer.

[16] Clearly if you knew exactly how the App software was structured, then it might well
be possible to predict something characteristic about the internal signal processing
such that you could confirm the software details from the dynamics of the scan-
evidence. The reverse engineer with this level of information at hand would have no
need to reverse engineer at all. He would simply pick the best software model and
implement it.

[17] Your old smart phone is not running the App (or any other program) as originally
written. Software is typically written in a human-friendly, so-called high-level language (e.g., using keywords like PRINT). When this program is loaded into a computer it is usually "compiled", i.e., automatically translated into a more

computer-efficient language (e.g., using explicit references to the computer's hardware such as registers and memory locations). It is (to gloss over a few more intricacies) this "machine-code" translation of your program that is actually being executed. Therefore, if you are going to find some way to extract the running program from the nanobot data, it will be this machine code program. This is the semantic model in terms of your smart phone's structure. It is functionally equivalent to the program as originally written, but it is not generally possible to derive the original program from this machine-code equivalent. The software as written in a high-level language is called the "source code" while its machine-code equivalent is the "object code". The transformation from source code to object code is quite straightforward (software called "compilers" do it all the time), but the reverse transformation is, in general, impossible, primarily because the compilation process reduces the general program to a machine-specific equivalent. Hence, software vendors typically sell the object-code version, which will be particular to your specific hardware device, of their product because the buyer can run this version but cannot understand it and cannot generate the original source code from it (which would permit all sorts of commercially undesirable machinations). Notice finally, that we have come up with two semantic models, a source-code one and a machine-code equivalent. So structure and meaning are not simple absolutes, but neither are they totally arbitrary.

[18] I am, of course, by no means the first person to promote the crucial essentiality of learning in AI. Way back in 1985, John Haugeland, in his refreshingly different AI book called *Artificial Intelligence: The Very Idea* (MIT Press), gave this possibility a small airing in a textbox entitled "Why not start with learning?" (p. 11). He cited two AI pioneers — John McCarthy and Marvin Minsky — who had both advocated the same view years earlier. As Haugeland says, a major stumbling block is that we first need to know exactly what should be learned; this takes the idea back to what is knowledge and how the human cognitive architecture manages it. These are necessary precursor problems that we are still a long way from solving.

ॐ Chapter 8 ॐ

Self-Organising Systems — The Engineer's Nightmare

"A few decades from now the forces unleashed by the bottom-up revolution may well dictate that we redefine intelligence itself, as computers begin to convincingly simulate the human capacity for open-ended learning."

Steven Johnson, 2001[1]

In the previous chapter I attempted to cut across all the bickering about how well or how badly the AI project — the quest to simulate intelligence on a computer — has done by offering a capability to learn as the acid test of an AI system. I also noted the obvious problems with this idea. Nevertheless, there is a good case for the centrality, even definitional status, of this aspect of intelligence — non-adaptive intelligence is an oxymoron.

You are constantly learning — a new fact (less than 10^{25} microseconds have elapsed since the Big Bang); a new possibility (the Big Bang never happened); a new concept (anti-matter); a new view of an old belief (the sainted Thomas More was happy to torture "heretics"); and so on. My examples, all rather grand, are not meant to be either an exhaustive list or a list of non-overlapping examples.

Notice, for example, that the second listed item has repercussions for the first: if you degrade your belief in the Big Bang as the starting point of the Universe, then you must consequently re-assess any facts about the Big Bang. That learning one thing may require reassessment (i.e. some degree of re-learning) of many other things has long been a huge stumbling block for those who attempt to build models of our knowledge.

The consequence of not "seeing through" all the possible repercussions of each new learning activity is, of course, the danger of developing inconsistent knowledge. For example, you're firm in your knowledge that less than 10^{25} microseconds have elapsed since the Big Bang, yet you become convinced that the Big Bang never happened. But which of us is rash enough to claim to be totally consistent in all our beliefs?

On the other hand our inconsistency is not usually rampant. In fact, it would be reasonable to claim that a general consistency in reasoning is a basic characteristic of human intelligence. So, like so many other aspects of our intelligence, the consistency of beliefs must be high but almost certainly not 100 percent. The role of compromise in intelligence is a topic to be re-opened when we get to the chapter on ultra-intelligence, and fully explored in the final chapter.

We all learn something every minute or two. For example, most conversations, even cursory exchanges, result in some modification of what you know or what you believe. You learn something, even if that something is to ignore the claims of your conversant. Arguably, a primary purpose of language is to convey information, and the consequence of the information transfer is usually to learn something. This is not necessarily a radical change to long-term memory. It might be as transient as a modified belief about whoever is saying things to you, such that you alter your response, then reassess your new belief and abandon it in the light of further exchanges.

I'll define "learning" as the capability of a system to change its structure, and hence behaviour, as a result of experience, and the changes must generally be for the "better". So the changes cannot be simply introduced by an outside agent, hence the synonyms "self-organisation" and "self-adaptation", which I shall use interchangeably to introduce a little variety.

There is now a considerable array of learning mechanisms that can, and have been, incorporated into software systems, at this point we must say, into AI systems. However, the creation of a self-adaptive mechanism is not in itself a passport to an intelligent system — some lead nowhere, some to dead ends, and others to chaos. These latter two possibilities, which may only emerge after some initial success, imply a further definitional rider: a learning mechanism must demonstrate its worth beyond a few initial successes.

Given this definition, we can acknowledge the current upper bound to machine-learning science: well into the new millennium, there is no machine-learning mechanism that offers any promise of leading to open-ended learning.

A learning mechanism, *per se*, is all too often treated as a clear road to intelligent systems (just over the horizon, of course). On the other hand, a machine-learning mechanism may be demonstrably no more than a yellow brick road to the disappointment of an Emerald City[2]. Unguarded enthusiasm is not usually voiced by the scientist who has developed the learning mechanism — he or she is only too well aware of the limitations and basic problems of their otherwise bright idea[3].

Let us consider one leading example. It derives from a "new" technology that we have invested some effort in already, and one that is typically held out as a light in the darkness of self-organising software technology — as indeed it is. It is neural computing where the well-lit potential sits next to the equally sharply defined scope and limitations. It is, as we shall see, only in the murk of ill-understood promise that the hoped-for and necessary potential can survive; when the light is turned on, the more shadowy hopes tend to disappear. Let's switch some light onto neural computing (and we'll further intensify the source in a later chapter).

The term neural computing covers a wide variety of distinct technologies — some self-organising, others not; some well-founded, others rather more cavalier; and some one-shot learners, others continuous. Elsewhere[4], I have laid out, and tried to organise this diverse landscape. Here and now, we just need to consider the big favourite.

So-called neural network computing is based upon computing with an array of simple (often identical) processing components/units. These units are interconnected by links that transfer signals from one unit to another, and often moderate the signal strength by means of a "weight" value associated with each link. An example is the Multilayer Perceptron network, the MLP, illustrated and explained in Chapter 6. It will be instructive to home in slowly on the barriers and breakthroughs that delivered this powerful technology.

In the 1960s a neural-network device that could be trained to recognise patterns, such as handwritten numerals, was announced and called the "Perceptron" (modesty and measured expectation in project nomenclature went out of the window right from the start[5]). Intermittently over the next decade, researchers struggled with trying to train these Perceptrons. They met with mixed success — sometimes the Perceptron learned the pattern-recognition tasks presented, and sometimes it did not. In 1969, Marvin Minsky and Seymour Papert published their results of a mathematical analysis of the Perceptron's capabilities. Their book *Perceptrons* clearly defined the scope of these devices — they were limited to learning only one sort of simple process[6]. This well-founded but unwelcome news put paid to research with Perceptrons. It also showed clearly the merits of formal analysis of learning systems as opposed to blind faith in an attractive but obscure technology.

Years of man-hours were probably wasted fiddling with Perceptrons that failed to learn their assigned tasks. The training regimes were modified and re-tried. The training data was changed. The Perceptron structures were modified. Many different combinations of these various changes were explored, often to no avail. In fact, there was no chance of success if the task to be learned was not sufficiently simple, but before Minsky's and Papert's analysis no one knew of this well-defined barrier.

Another decade on, and the lessons had been learnt. A neural-network device called the Multilayer Perceptron (MLP) was launched together with its training process called the "back-propagation of error algorithm", or more manageably the "Backprop Algorithm". This combination is well-founded in that the Backprop learning algorithm is guaranteed *never* to degrade the MLP's performance. As we saw earlier, this machine-learning procedure automatically adjusts all the network link weights on the basis of feedback of the error between the output signals the network generates and the correct output signals for the input signals applied. In addition, the so-called "hidden layers" in these networks (in our example in Chapter 6, there are two hidden layers: one composed of four units and the other composed of two) mean that they have the capability to learn any task (for which sufficient input-output pairs are available). The Perceptron limitation was lifted.

Further good news was that after a few trial runs (to sort out such things as how many units to have in the hidden layer, or layers), an MLP for a chosen application is easy to construct. All units in each layer are connected with random weights to all units in the next layer — input to hidden, hidden to next hidden, and finally, hidden to output. The need for a set of data samples for training introduces one constraint — there must be a way to get hold of a set of pairs of input signals each coupled with its correct output signal, a sufficient number of input-output pairs. This requirement, of course, scuppers any hope of using MLPs for open-ended learning. Typically hundreds up to tens of thousands of such data pairs are required; it depends upon the complexity of the process to be learned.

Skating right over the fine details, this network can be automatically trained to compute the function exemplified by the set of training data samples, the input-output pairs. The target might, for example, be that of learning to recognise a variety of handwritten characters. In which case, the training set will consist of many different examples these handwritten characters each paired with its correct identification. What the learning process amounts to is repeated fine adjustments to the link weight values, which is done automatically by means of hundreds or thousands of small changes by the Backprop Algorithm.

199

A vital element of the well-foundedness of MLPs is the guarantee that the Backprop Algorithm will never adjust the link weights to make the network worse. A second is that, in principle, MLP learning is not restricted to the simple tasks that so limited Perceptrons.

In summary: for this type of neural-network computing the learning procedure is one of changing the weights, the strengths, of the links between the processing units. And this procedure is intended to give a new version of the neural network, a version that is a "better" system, e.g., the new version makes fewer errors (or smaller errors) than the previous version. In the above example of character recognition, a "better" network will correctly identify more of the training characters, or get closer to correctly identifying more of them[7].

Without such a solidly based technology (such as we lack with most other neural computing, and almost all machine-learning ideas), attempted learning may result in an endless chase around with the network getting better, then getting worse again, and so on. So the technologist using most other neural computing technologies doesn't know if the network is going to improve when learning, and doesn't even know if the network is capable of learning the task being attempted.

When a non-MLP network fails to learn, it's an open question whether the training regime, the training set, the network architecture, the process to be learned, or any combination of these four individually complex factors is to blame. "Try something and see what happens" is the best, but hopeless, strategy, and not one that sits comfortably within science or practical engineering. It is thus not at all surprising that MLPs dominate this field, and have done so ever since their invention.

With all this experience and experimentation, has MLP technology been improved over the intervening decades? Yes, there has been an accumulation of useful practical knowledge about setting up initial networks efficiently, and there have been improvements to training with better, quicker versions of the Backprop Algorithm. But fundamentally, nothing has changed.

We can generally train MLPs to do a pretty good job of learning many different processes, but each trained MLP will compute just one process and after training, the link weights are generally fixed and learning ceases. Why? Because if properly done in the first place the chosen network architecture with the available training data was trained to an optimal point. It can't get any better unless the nature of the input signals changes, in which case the Backprop Algorithm may be re-employed to re-optimise the network.

Because of the well-foundedness of MLP technology, it is a valuable and widely used one. And also because of this well-understood basis, we know that MLP technology is no more than a slick and powerful way to optimise a process, but only a process for which we have access to a good many samples of input-output pairs, the necessary training and test data.

Hawkins, who is developing a cortex-based theory of intelligence[8], is right in his dismissal of neural-network technologies, and MLPs in particular, as a computational basis for intelligence. This leads him on to search for something better which he claims to have found in the neural organisation of the neocortex — his "memory-prediction" theory.

His mistake, I believe, is to abandon the constraints of formal analysis, and to simply collect together structures and processes that apparently serve his purposes. This is yet another theory whose scope and limitations will be anybody's guess (even if his 11 "testable predictions" do permit unequivocal testing).

Is this MLP technology going to lead to anything that vaguely approaches the flexibility and sophistication of human learning? No.

Are any of the other self-organisation technologies that we know of? Maybe, but not before a well-founded understanding of these technologies has been gained, and we discover how to extend them.

Ill-understood self-organisation technologies must harbour mysteries. The casual observer may see these mysteries as hiding the potential to solve some fundamental problems. But the technologist with experience of grappling with the ill-understood sees only an unmanageable technology threatening emergent chaos; the mysteries usually hide only unpleasant surprises. Maybe one day a poorly-understood technique for self-organisation will surprise us all and deliver open-ended, human-style learning. But, like waiting for Godot, it could be a very long wait, and can never be a viable basis for the next big breakthrough in AI.

A good example here, and one that the Singularity programme explicitly endorses, is Hebb's suggestions, at the dawning of AI, for human learning as "reverberating" "cell assemblies" of neurons[9]. Through repeated use and hence stimulation, clusters of neurons (his "cell assemblies") begin to mutually stimulate each other and so produce a "reverberating" substructure — a group of neurons that can self-sustain a state of excitation for a short time. This idea presents perhaps a neuronal mechanism for "short-term" memory — e.g., when trying to memorise a friend's telephone number mental rehearsal can help, but the "memory" needs to be used immediately (or preserved by writing it down) before it fairly quickly decays away.

In 1980, I joined forces with several other Hebb fans to develop the ideas of Hebbian learning[10] in a computer model. We soon found that we had to make so many more or less arbitrary decisions (such as how much activation and for how many time steps leads to "reverberation" which is then decayed at what rate), that the behaviour of the working model was determined by our (arbitrary) decisions rather than by Hebb's interesting ideas. "What fires together, wires together" may be right on as a useful slogan that is borne out by research into human learning at the neuronal level, but it is a far from a useful basis for model building.

Designing and developing large software systems, IT systems, is a process that pushes hard on the limits of human comprehension. Grappling with understanding the complexities of very large programs always has software engineers working well outside their comfort zones because programming technology of all stripes demands an excessive attention to fine detail, and every detail is potentially crucial. Witness the many budget busts, deadline overruns, and subsequent failures to work properly (which have led to the abandonment of numerous very expensive and time-consuming IT projects)[11]. In May 2013, for example, the UK's BBC announced the abandonment, after five years of effort, of a project to digitise all their output in a readily usable form; £100 million had been wasted.

One of the very few saving graces of a modern IT-system's structure is that it remains fixed; it doesn't change from what was originally programmed, it doesn't re-organise itself — it doesn't learn.

One thing that the software engineer would like to be sure of when struggling to get a conceptual grasp of an IT system (whether he originally wrote it, or not) is that the program *as written has not changed*. The original programming was presumably based on an explicit design with certain (hopefully, also well documented) goals, and was tested thoroughly. But if that's all changed because the system has learned from its experiences, the software engineer's already unmanageable task has escalated. The difficult job of interpreting somebody's best efforts has escalated by the additional need to also interpret a learning algorithm's actions. This learning process will have introduced changes as a result of the details of the system's history. What changes have been made, and how exactly they have been introduced is dependent on the details of the learning algorithms. Unravelling this degree of indirect consequences in an IT-system's innards threatens to... to what? Words fail me.

Open-ended learning sounds great as a mechanism for supporting intelligent program behaviour; but it's going to be a nightmare for any software engineer who has to understand a self-re-organised program because he/she is required to correct an error, or modify its behaviour. Of course, if this open-ended learning is introduced right at the end of the grand Singularity-system project, and if no further human intervention will ever be needed, then perhaps it could be viewed as a positive addition (provided that we know for sure that it will not run riot). But if learning is, as I have argued, integral to intelligence, it can't possibly be a final add-on component, can it? It is even less likely to be added on correctly at the first attempt.

Central to the problem of discovering appropriate and sufficiently powerful learning mechanisms is that of "knowledge" which is also a basic requirement of intelligence (even the newborn child "knows" something?). The goal of the learning is to rearrange and extend knowledge such that the intelligence that works off it is similarly improved, or at least maintained at its current level in a changeable environment. What is this knowledge and how does it fit into the cognitive architecture?

At this point you should not be at all surprised when I tell you that these questions have no clear answers. You may recall from Chapter 1 the "toddler robot", iCub, that "learned". Apart from asking what learning procedure it uses, and how well-founded this procedure is, we still do not have quite enough information to properly probe iCub's behaviour. We need, as a first step, to get to grips with "knowledge". It is time to look into this "knowledge" that you and I both posses and draw on constantly to support our intelligent behaviour.

Endnotes

[1] Steven Johnson's *Emergence* (Penguin, 2001), p. 208. The book is a very readable account of "emergent" phenomena. It is interesting enough on ants and cities, but then the usual journalistic misleading glosses and wildly over-enthusiastic predictions

for software systems and particularly the "new" technologies, such as neural networks, spoil the read for anyone who can see beneath the labels.

[2] It is true (in the story at least) that Dorothy and her strange pals all found what they were looking for in the Emerald City (if not in quite the grandiose forms expected), but the Wizard was a clear fake.

[3] There are exceptions: Kevin Warwick, for example, a Professor of Cybernetics, whilst being interviewed (BBC Radio 4 *The Choice* on Tuesday June 14th 2011) about his decision to have an implant to open two-way communication with a computer, conned the interviewer Michael Buerk. Kevin Warwick persuaded Michael Buerk that scepticism about intelligent machines was misplaced because machine learning is now a technological reality. Warwick laughingly put his interviewer's outdated view into a modern perspective by saying, "A lot of machines actually learn and adapt." Well, it's true, some do. The reality, however, is far short of the adaptability that all animate life possesses (and arguably inferior to even the lowly bean seed that can sort out "up" and "down" for leaves and roots respectively, and also negotiate all sorts of obstacles that may be in its way).

[4] In *A New Guide to Artificial Intelligence* (originally published by Ablex: NY in 1991, and now by Intellect: Bristol) I attempted to organise and illustrate the full sweep of neural-computing technologies. A couple of decades on, I can detect no necessity for significant reorganisation, or extended characterisation, of this exposition. The multilayer perceptron (MLP) still dominates; its learning algorithm (the backpropagation of error, the BP algorithm) is fundamentally unchanged although arguably significantly more efficient.

[5] In the euphoria of the first blossomings of AI, many projects were similarly unfettered by any suggestions of modesty, for example, "The General Problem Solver, a program that simulates human thought" by Allen Newell and Herbert Simon in *Computers and Thought* edited by E. A. Feigenbaum & J. Feldman, pages 279–293 (McGraw-Hill, 1963).

[6] *Perceptrons* (MIT Press, 1969) by Marvin Minsky & Seymour Papert presented a mathematical analysis that revealed that Perceptron networks were only capable of learning "linearly-separable" functions. Roughly put: plot the values of a two-class function on a graph with the usual x and y axes, then if values of the two classes can be separated by a straight line (as opposed, say, to a wriggly one) then the function is linearly-separable. Although defining linear separability and determining if a function is linearly-separable are not simple processes, it is quite clear that many simple functions are not linearly separable, and thus a Perceptron can never learn them.

[7] The raw output signals (i.e., before any thresholding) from an MLP will be values between 0.0 and 1.0, so if the correct signal (for a given training sample) is 1.0 then an output of 0.9 is "better" than an output of 0.5 although neither is correct.

[8] *On Intelligence* by Jeff Hawkins with Sandra Blakeslee (Henry Holt: NY, 2005).

[9] In 1949, Donald Hebb published *The Organization of Behavior* (Wiley: NY), an interesting and stimulating ramble over a variety of possibilities for mechanisms that

might contribute to cognitive architecture. Hebbian learning has always attracted a following. Its basic notion of repeated use of a connection resulting in enhanced facilitation of that connection is superficially attractive (and biologically plausible) but quickly reduces to inspired guesswork once the researcher attempts to theorize seriously and model Hebb's ideas.

[10] The project results were published in 1983 in *The Journal of Theoretical Biology*, vol. 100, pages 485–509, under the title "A Neural Theory of Cognitive Development" by V.S. Johnston, D. Partridge and P.D. Lopez.

[11] Wherever you look in any society with a well-developed software-engineering infrastructure (such as the UK and the USA), you will find it littered with IT systems that have eaten up multi-million dollar/pound budgets and years of effort yet have either been abandoned or fail to work properly. My book *The Seductive Computer* (Springer, 2011) is focussed on this IT-system crisis, and places the blame squarely on the discrete technology of modern programming and the excessive demands it places upon its users. And this inevitable unmanageability all occurs in the absence of the further complexity escalation that self-adaptive mechanisms would introduce.

ರ Chapter 9 ಙ

The Knowledge Web

*"If you can solve the problem of search, that means you can answer
any question. Which means you can do basically anything."*
Larry Page, 2002[1]

*"Machines have exacting memories. Contemporary computers
can master billions of facts accurately."*
Ray Kurzweil, 2005[2]

"The more you know, the slower you go."[3]

Knowledge in the form of facts can easily be stored in computers, and facts also offer the possibility of being mastered, but these are nothing like the same thing. Storage is easy; it is however mastery that is a likely key to intelligence. If we replace "master" with "store" in the second opening quotation, I'd happily go along with it. But as it stands, it's just plain wrong (given any vaguely reasonable meaning for "master"). So what are these facts, and what sort of mastering is needed within an intelligent system?

It appears self-evident that any cognitive architecture that will be up to the job must contain a facility for storing and retrieving learned facts as well as many other things, such as, say, Pythagoras' rule. These stored items will contribute to the system's knowledge as will combinations of the two — i.e., a general rule applied to facts in order to derive new facts. So, I might, for example, store the fact that the square of 10 (i.e., 10×10) is 100, but use my stored multiplication rule (together with a pencil and the back of an envelope) to derive the square of 16.

You know a lot of things about the world — perhaps not grand, scientific things necessarily, but very many things nevertheless. You know that night follows day, that if you (or anyone else) venture out in the rain you will get wet, that Barack Obama is (or was) the first African-American President of the USA, that books are for reading, that reading requires knowledge of the book's language, and so on.

How many facts have you learned? There would seem to be no end to what you can list, but an infinite memory capacity you do not have. In our finite brains none of us do. Perhaps this apparent conundrum is resolved by what we call a "mastered" fact and by what further facts we can generate if we try.

So, I might justifiably claim to know an infinite number of numbers, but this doesn't mean that they are all stored in my brain; I just know the rule that allows me to keep counting — 1, 2, 3, 4, etc. — forever, in principle at least. It is true that once in the realms of billions and trillions it will not be so easy to keep saying the next number, and, of course, "forever" is a good deal longer than my anticipated life expectancy.

Quite apart from the meaning of "mastery" of facts, there is the knotty prior problem of what exactly is the roll call of facts that are needed to support intelligent behaviour — what is a fact? And where do the rules, such as "add one to any number to get another, bigger number", fit into your knowledge framework?

Knowledge itself is sometimes described as "justified true beliefs". Comforting as such a description is, with this as our definition we cannot be said to have a great deal of knowledge. The numbers and their manipulations might still be secure knowledge (at least until some bright spark spots a basic flaw in number theory). It seems to me that, on the whole, certainty is just not within our reach. Rather, we must act on the basis of "well supported beliefs" if possible, and on "likely beliefs" or even "plausible beliefs" otherwise. So there is a further twist to the knowledge puzzle: whatever we store or master, we must also manage a system of associated confidence in the truth of the various elements of our knowledge because very little of our knowledge is absolutely true.

Your knowledge is clearly much more than a list of facts. It must involve rules for generating further facts. This collection of facts and rules must be organised for efficient use. You clearly cannot search through everything you know whenever a piece of knowledge is needed, can you? Relevant knowledge just jumps into consciousness. Well, not always, but sufficiently often to be persuasive of a degree of interconnectivity between elements of your knowledge. Your knowledge is not just a bag of independent facts. The now familiar issue of modularity versus global accessibility re-arises as encapsulated specialised knowledge versus widely available knowledge. In addition, there must be a system of belief management associated with all your knowledge.

Science viewed as "organised doubt" is a description that puts clear distance between certainty and proof (those popular, but generally erroneous, attributes), and the reality of what the scientist unearths. The wealth of knowledge upon which your intelligent behaviour appears to be founded, might be similarly described — with reduced emphasis on the "organised". As we touched upon in the previous chapter, much of the learning that is integral to intelligent behaviour is not the addition of new facts or rules (nor their abandonment). It is modification of beliefs about the facts and rules, modification of the belief structure that shadows the facts and rules.

Since Lashley's fruitless search for "the engram" in the late 1940s — a search for whereabouts in a mouse's brain the memory of how to traverse a maze was stored[4] — our understanding of the architectures of cognition has progressed beyond his vision of the brain as a homogeneous mass of undifferentiated, equipotential regions. With respect to knowledge, however, our understanding has barely gone beyond box diagrams for long-term and perhaps short-term memory locations.

For example, it is pretty well accepted that there is a brain region specialised for the recognition of faces and that a distinct part of the brain (called the hippocampus) plays an important role in consolidating memories so that they can become permanent ones. But where, how and if we store such facts in our brains remains largely a mystery (although all sorts of research into the possibilities is "well underway"). Whereas where and how we manage the associated belief structure barely merits even the dubious epithet of "hopeware".

Psychological experiments reveal a variety of clues about how knowledge is stored and how it is accessed. A division into, for example, short-term and long-term memory was long assumed to be a fundamental partitioning. Your short-term memory capability will be limited to just a very few items — famously "7 plus or minus 2". It is just this number of items that you can "keep in mind" as long as you continue to rehearse them. So when given the telephone number of an old friend, and a pencil is not at hand, you repeat the number to yourself while searching for the pencil. If a pencil is not forthcoming, and you start to think about something else, then in all likelihood the memory of the telephone number will be gone, or at least doubtfully remembered.

Alternatively, if this exchange of information happens in, say, the middle of a swimming pool and so you know that the necessary pencil is not to be had in the near future, then you might focus on remembering the number in a more permanent manner. You might for example realise that the number starts out the same as a friend's street address and that is followed by your cash-card pin number in reverse. These realisations will enable you to "tie in" this new fact with your vast store of long-term memory knowledge. It is this "tying in" that may well be the essence of "mastery" as distinct from mere "storage".

As a result of such conscious forging of links there is a good chance that you will establish a relatively permanent memory of the new telephone number. You will have put it into your long-term memory such that when you get home you can retrieve the new number and write it in your address book.

Curiously, the fact that the "tie ins", the associations that will allow you to retrieve, or reconstruct, the new memory are somewhat odd (best friend's street address and reverse pin number) appears to be a significant feature of successful conscious effort to memorise.

However, beyond drawing boxes with interconnecting arrows, one box with about seven divisions labelled "Short-Term Memory" and another, effectively boundless, labelled "Long-Term Memory", we have not made a great deal of progress in our search for how knowledge is organised in support of intelligence. But psychologists have, of course, gone further than this very simple characterisation.

There has, for example, been a real proliferation of Long-Term Memories: *semantic* (to account for the facts — e.g., that the pool-centred telephone number is that of a long-lost friend), *episodic* (e.g., that you were given the number in the pool and, in the absence of a handy pencil, you "worked" to tie it into your memory), and *procedural* (e.g., that you know how to swim) — to name just three.

Rather than proliferation, Short-Term Memory has morphed into Working Memory, a concept that moves beyond "7 plus or minus 2" chunks to emphasise the importance of process as well as capacity. Conceptions of Working Memory have included: a *phonological loop* that refreshes items in a phonological store (like repeating a telephone number until you can write it down), and a *visuospatial sketch pad* for visual images with these two components controlled by a Central Executive that plans, monitors, sequences, transfers, transforms, etc. One outgrowth of this was the development of measures of Working Memory Capacity, and so it goes on, but I will not[5].

Because most scientists whose efforts are focussed on understanding intelligence do not work from brain structures, we must ask: what is the route to their insights into what's going on in our heads? The answer, of course, is experiments that manipulate our behaviour in clever ways and so directly gather evidence of the working of the mind. Indeed, studies that employ brain imaging are guided by the box-and-arrow models — models that are developed on the basis of the type of mind-experiments we are about to look at.

The vast scope of experimental psychology and cognitive science is way beyond the aspirations of this book, but we will touch on a few experiments that are well within the capacity of every reader to perform. The particular experiments chosen will also be ones that shed some light on our current focus — knowledge.

This will be no more than a tiny dip into the oceans of experiments and data, and it will also ride roughshod over the many subtleties and controls necessary to generate good quality "scientific" evidence. We will, for example, experiment only with you and me, whereas a proper experiment would need to be repeated many times, and with many different persons, in order to get reliable general results and not risk the mistake of assuming that you or I have brains that are totally representative of the norm in all respects.

To begin with, you can play the free-association game. Look at this word — DOCTOR — and then note down the first new thoughts that emerge. Indeed, you will find it impossible not to make these follow-on associations, which begins to suggest that the knowledge in our heads is a network of interrelated things: words, facts, concepts, ideas, images, etc.

What results do you get? NURSE, HOSPITAL, ILLNESS and INJURY are good bets, and COMPACT DISK, OAK TREE and BANANA are not (unless, of course, you happen to know a Dr. Banana).

Repetition of such word-association experiments with many different people will deliver counts of the number of times each response to a given stimulus word, such as DOCTOR, is encountered. So after 100 tests, we might find that the first response word was NURSE 40 times, HOSPITAL and ILLNESS both 20 times each, INJURY 10 times with COMPACT DISK, OAK TREE, BANANA and seven others all once only — that's the full 100 responses taken care of, is it not?

A conclusion that may be drawn from these experiments is that NURSE (the word, or the concept — can they be separated?) is closely associated in our minds with DOCTOR which is not too surprising. HOSPITAL and ILLNESS are also quite strongly associated to DOCTOR, but not as strong an association as NURSE, and so on.

By such means we have learned something about the knowledge in our memory structures, but what? That our word/concept memories may be organised by means of direct associations whose strengths differ in relation to the number of similarities between them. So DOCTOR and NURSE have a whole host of similarities — both are persons, treat the sick and injured, work in hospitals, often marry each other, and so on. DOCTOR and HOSPITAL also have a number of similarities, but not as many as DOCTOR–NURSE. DOCTOR and HOSPITAL are, for example, different sorts of things, one is a person and the other is a building.

You can play this game for as long as you wish, and there is a plausibility about my "number of similarities" analysis, but we know that science based on plausibility arguments is on a par with Greeks bearing gifts (or promising austerity measures).

Suppose, for example, that you are the spouse of a doctor who has just run off with a nurse, then your first free association to the word DOCTOR might be BASTARD, and your second might be that too. If so, then do we have evidence that personal history (as well as timing) can override associational strength based on the number of abstract similarities?

Whatever the reasons behind the differential frequencies of response, this phenomenon does seem to point to an associative model of memory (whatever that quite means), and one in which association strengths vary quite consistently among the individuals of a given culture. In addition, dynamic flexibility emerges as an important feature: so the usual DOCTOR–NURSE strong association can, in an instant, be replaced by DOCTOR–BASTARD, which over time degrades to DOCTOR–EX-SPOUSE with an associated mood change from fury to gratefulness.

This flexible associative memory structure appears to be quite different from a conventional computer database with predefined access keys — i.e., the usual method for storing facts in a computer system. The database of UK vehicle licences, for example, is set up to allow the operator to give it the details of a licence plate (a "key") and it will retrieve the owner of the vehicle, vehicle type, etc. (the related "facts"). It might also be structured to accept a street address (another acceptable "key") and then retrieve the vehicles registered at that address. It will not, however, accept a breed of dog (not a designed-in "key") and retrieve all vehicles owned by persons who have a pet of that type.

You, too, may be pushed to retrieve the relevant knowledge when confronted with such an odd request, but given the dog-breed example, you may well be able to deal with it sensibly. You may even be able to come up with one or more vehicles whose owners have pets of the specified breed, or you may just conclude that you do not know of any such vehicles.

The point being that your knowledge does not appear to be structured in any hard and fixed, predefined manner. There is an inherent flexibility in human knowledge that computer databases, despite their speed and accuracy, do not exhibit. So how do you do it? We can take another small step into the world of the cognitive scientist probing into this mystery.

The earlier word-association experiments allow the psychologist to determine the closeness of association in our minds of certain word pairs. In particular, certain pairs can be classed as strongly associated, others less strongly associated, and yet others very weakly associated.

Now imagine yourself seated in front of a computer screen (not too tricky I trust). You are told that a short string of letters will suddenly appear on the screen and your task is to decide whether it is a word of English or not. On the keyboard are just two keys labelled "WORD" and "NON-WORD", and you answer by pressing the appropriate key as quickly as you can according to the decision you make.

What the experimenters will measure is the time taken to deliver your answer (i.e., time between appearance of the letters on the screen and your answering key press), the "response time". They will also note whether you get the answer correct or incorrect. These two measurements will constitute the evidence collected.

Here's a possible first word:

DOCTOR

You make a judgement, word or non-word, and push the corresponding key, then the next stimulus appears:

NURSE

Again you make a decision and depress the appropriate key.

So how are the "response times" and "correctness" evidence pertinent to knowledge organisation?

You are almost certain to respond very quickly with the correct answers on both occasions, which are WORD and WORD (in that order). Not very enlightening you might think, and you would be correct, but the experiment continues with various other pairs of stimuli until at some point you get:

SAILOR

Followed by:

NURSE

And again you are very likely to correctly conclude that these are both WORDs. But the telling difference between this pair of stimuli and the initial two is that your response time to NURSE will be slower in this latter case (such measures have to be averaged over many tests and many people).

This sort of result, which is reconfirmed with many different word pairs, is taken to show that if the second word is closely related to the first word (such as DOCTOR–NURSE) then you can recognise it as a word more quickly than if the two words are unrelated (such as SAILOR–NURSE). In addition, by testing different degrees of relatedness (e.g., HOSPITAL–NURSE might be an intermediately related pair), correspondingly varying response times are found.

This effect suggests that your knowledge of words is not just a list arranged, say, alphabetically; it is further support for the idea that your memory is a network of associations of varying strengths — a web of associated words, or concepts. And moreover, because the process of recognising a word is quicker if it's closely related to your previous word recognition, there is either some "priming" activity that carries over from one word recognition task to the next, or perhaps "stronger" associations that are traversed more quickly (or both).

Further information is elicited by word pairs such as DOCTOR–NERSE and SAILOR–NERSE. In both cases the second decision should be NON-WORD, and what is consistently found is that this decision is more often incorrect following DOCTOR than following SAILOR. We also tend to take longer to decide that NERSE is not an English word when it follows DOCTOR than when it follows SAILOR.

This is, for us, new information pertaining to the structure of human knowledge, but what exactly does it tell us? Is it a natural consequence of the rush to judgement resulting from the very close association between DOCTOR and NURSE (which, of course, sounds just like

NERSE in your head)? For strongly associated words the "spell checking" may be glossed over and so necessitates "backtracking" when the NON-WORD realisation dawns. The cognitive scientist, who would surely probe around much more extensively before drawing any conclusions, might conjecture that a stronger link is traversed more quickly and cavalierly (than a weaker one) so then any need for subsequent "spell checking" means a step backwards, and that takes longer.

Such an analysis might well account for all the data, but does it give the scientists a useful model of knowledge organisation? Not really — it just gives a few clues about how we store and retrieve some of our knowledge. It can provide the scientist with some constraints on what modelling structures to use, but that is a far cry from anything close to complete guidelines for model building.

For the model builder, there are, for example, many different ways to construct a process that "spell checks" potential words with differing degrees of exhaustiveness to be determined by the association strength from some stimulus word (such as DOCTOR) to a target word (such as NURSE).

Here's the picture that tells us something about theories of memory organisation but also exposes some of the many mysteries remaining:

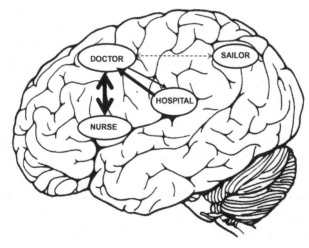

In this link-and-node network, which is an extremely cursory example of a once-favoured structure known as a *semantic network*. There are many unanswered (and mostly unanswerable) questions about both the links and the nodes (but apart from those, no problem). For example, should there be a lozenge labelled "bastard" in order to accommodate the doctor who's deserted her spouse? Is "bastard" a similar sort of concept to, say, "hospital"? They are both words that might spring to mind in similar circumstances, but one is a solid object universally associated with doctors, and the other is more of a transitory, situation-specific, descriptive term.

I illustrate the strength of association in terms of thickness of the arrow links. Some are two-headed, showing that the same closeness of association holds in both directions, and others are not. But can these association strengths really be fixed features (even neglecting change over time) when they apparently vary with context?

Because you have absolutely no trouble understanding the sentence, "The nurse attended to the sailor in hospital.", we might infer that I've omitted the links that must exist between the NURSE, HOSPITAL and SAILOR nodes. Is every node explicitly linked to every other node? Maybe NURSE and SAILOR are only linked indirectly via, say, PATIENT and PERSON?

What do my nodes represent? Is the DOCTOR node my concept of "doctor" or simply the word "doctor", or something in between, or something else? Part of our earlier analysis of experimental results involved "spell checking" of the word itself, so that must be in the node surely? But could that be all? It seems unlikely given the flood of information that cascades into consciousness when you see the word DOCTOR. In addition, most of us know and use many fact and concept words that we cannot spell correctly.

One commonly held scientific view is that a word is represented as an interconnected threesome: (1) the spelling that we use when we read or write it; (2) the sounds involved in our listening and speaking of it;

and (3) what it means. So, for example, your knowledge of the word BANK, which can be either a RIVER BANK or a FINANCIAL BANK, would have two different meanings attached but only one each of the spelling and pronunciation components. A word such as BASS, which can be a tasty fish or a low musical note, would have a single spelling connected to two pronunciations and two meanings (which must be correctly paired).

With just these rudimentary hypotheses about what knowledge we learn, and how we might store it, we can begin to probe the learning abilities of the "toddler robot" iCub. You may recall from Chapter 1 that iCub "learned" two objects: an "octopus" and a "purple car".

So what information might iCub have learned? Following the "three-some" hypothesis we might expect the following new knowledge after the learning had taken place:

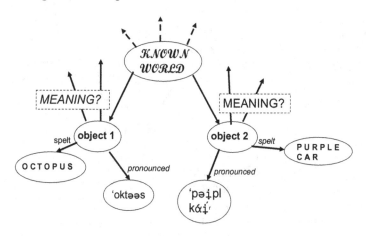

In this diagram of iCub's learned knowledge, which is pure guess-work, we see "object 1" which is the one presented as "octopus", and "object 2" which is the one presented as "purple car". These two new pieces of knowledge are linked to the pool of everything iCub knows, signified by the KNOWN WORLD node, which also links out to other things that iCub presumably knows (but we don't know what they are).

The "spelt" link from each of the learned objects details how the name of the object is spelt, and the "pronounced" link indicates how each is said (I've adopted the phonetics of the Collins dictionaries which can be treated as a general mystery tempered by the assurance that a computer system could transform such hieroglyphics into sound and vice versa).

It's the third element of the threesome that's really problematic. It's all very well to state that there must be a "meaning" component to any learned knowledge. Few would challenge this assertion. But what does such a "meaning" component amount to?

Here's a cursory stab at a possible answer:

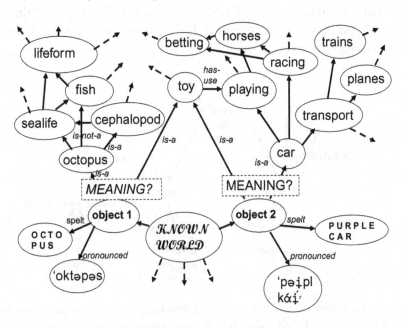

Such a web of "meaning" has some plausibility, does it not? It seems to capture, for example, the knowledge that both objects are toys. So we might ask iCub, "Is the octopus a toy?" In its computer memory it could follow the "is-a" link from object 1, the octopus, to the node labelled "toy", and answer, "Yes, the octopus is a toy."

Although this severely limited attempt at the meaning web has a superficial plausibility, as an integral part of learning the two objects it also leaves a lot to be desired. You can quickly identify countless omissions and general difficulties with it.

So, for example, if we asked, "What colour is the purple car?" iCub would be stymied, on at least two counts. Firstly, it appears (from this representation of its knowledge) to know nothing about colours in general. Secondly it knows nothing about the colour of the purple car in particular because for iCub "purple car" is simply the object's name, no different from, say, "Ford car" or "small car". Indeed, it is probable that iCub would just as easily "learn the red banana" when presented with the purple toy car. Why? The difficulty goes way beyond colour ignorance.

Within all the discussion of "meaning" I have presumed that iCub does extract some meaning from the objects it "learns". How might it do this? Just like you and me it would have to link the name component "car" with its knowledge of cars, and it would have to visually analyse the object and extract further links into its general knowledge, e.g., that the car was a toy, not a real car. I seriously doubt that it does any such analyses, in which case the "meaning" component will be devoid of content.

These inadequacies puzzle you and me because, for example, as sophisticated colour-knowledge users, we cannot help but "know" the colour of the "purple car"[6] as well as know that "colour" refers to a property of the object. Furthermore, we can check visually that the car is purple in colour. "No problem," the iCub researcher might say. "We can soon add colour knowledge." We do know that iCub must have knowledge about its position in space because it successfully touches the purple car.

So here's a realistically plausible knowledge representation that answers the "colour objection," and is adequate for the demonstration presented.

221

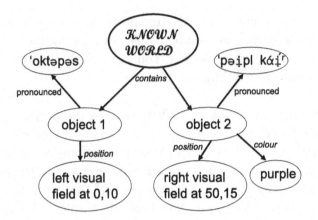

With this representation of its knowledge iCub could tell us the colour of the purple car. But it couldn't tell the colour of the octopus or whether the objects are toys. Okay, so we'll add a few more nodes and relationships, say, the node "toy" with an "is-a" relationship on the links from both objects, and so it goes on.

The point being (and this is a fundamental characteristic of virtually all AI demonstrations) that all necessary knowledge can be programmed into iCub's memory. Provided we specify exactly what knowledge is required, iCub can be programmed to "learn" it. It can also be programmed to "master" both the octopus and the purple car provided we pre-specify exactly what we mean by "master".

It's the need for unanticipated meanings that cause the problems — crass blunders, or mystification, or system failure. Any of which, with repetition, signal that intelligence has not been achieved. The knowledge representation gets larger and larger as a consequence of fixing up the system's inadequacies when each one comes to light as a result of experimentation. It also gets more and more complex because of the need to link new pieces of information into the total knowledge web. And so it gets progressively more difficult to integrate new knowledge without introducing any undesired consequences. This is precisely why so many AI projects take off at high speed, support impressive initial demonstrations, and then slowly but surely get bogged down in their own complexity.

A source of complexity, exemplified by this iCub example, derives from general knowledge of octopi — e.g., they live in the sea — and the specific restrictions that apply to a toy octopus — e.g., it doesn't live in the sea. This particular trickiness leads to an inconsistency that is already present within our first representation of iCub's knowledge: just follow the two "is-a" links from "object 1".

Within iCub's knowledge there must be a mechanism for cancelling much of the usually pertinent general knowledge when the object is a toy. And for you and me countless different cancellation structures and extensions will be necessary to account for our mastery of dead octopi, drawings of octopi, eating octopi, and so on.

Notice also that so far, we've been mixing general knowledge and knowledge that is particular to these two objects. So all octopi are cephalopods, but this particular one is strangely blue in colour and has been placed in a specific location within iCub's visual field. Is there a case for distinguishing the relatively timeless general knowledge from the transient (perhaps fleeting) knowledge of what is immediately occupying our attention? Is this another angle on long-term and short-term memory?

Somewhat more down to earth, we have no idea what iCub's knowledge representation is, nor what it knows beyond the two objects it "learns" in the demonstration. As a minimum, its knowledge after the double learning exercise might be as shown on the next page.

This knowledge representation makes it clear that iCub knows only the two objects it has just learned, and it only knows two things about each. Firstly, it knows the sound of the name of each object which I illustrate as the cryptic phonetics in each node at the end of a link labelled "pronounced". Secondly, it knows whereabouts each object is located within its visual field.

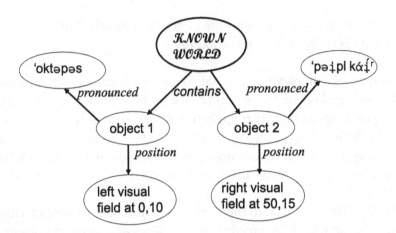

Precisely these, and only these, two pieces of knowledge are required for the task we see it perform (actually it only needs to know where the purple car is). Yet we are strongly inclined to "see" intelligence behind the performance, a demonstration that could be supported by the simplest of knowledge. How should the objective scientist counter this subjective rush to judgement — perhaps a deep genetic urge?

Echoing the style of the Turing Test, iCub's demonstration must be extended and under the control of the investigating scientist rather than a carefully choreographed short demonstration by the system's designers. Given that no robot approaching the language competence nor the manual dexterity of a two-year-old child is yet available, the system designers would need to circumscribe explicitly the nature of their system's intelligence competences: what it "knows"; what it can learn; what questions it can understand; what actions it can exhibit, etc.

Such a description would constitute a framework within which the enquiring scientist can probe the system's capabilities. And as we now know, it is the *failure* of an experiment that makes a solid contribution to the understanding of the "world" being investigated. Recall the Eleusis card-game examples and the derived slogan:

Successful experiments are most pleasing, but unsuccessful ones are most informative.

However, a series of stunningly informative failed experiments with a robot may be great for science, but makes for poor entertainment. It is also fundamentally unsatisfying for the human psyche — the designers get no joy from witnessing the failures of their brain-child. The audience, although much less committed to this particular project, would nevertheless rather see some impressive little cameos. And the impartial investigator is probably not delighted by a succession of failures to induce the robot to do something interesting. In addition, the wealth of scientific information gained by a series of experimental failures is pushing against our deep-seated predisposition to "see" intelligence lurking behind the scenes.

The iCub is verbally instructed to "touch the purple car" which it proceeds to do. The verbal instruction (rather than, say, typed in) means that iCub must be able to interpret the sound patterns of the object's name. This not an easy task. It would appear that it has been simplified, as it usually is in speech recognition systems, by training the system to the pronunciation of its instructor[7]. In addition note that, as far as the demonstration goes, iCub has no need to understand the full phonetic alphabet; it just has to be able to distinguish the sounds for "purple car" from those for "octopus" because they are the only two objects it must decide between. We'll delve into iCub's language competence in a subsequent chapter.

This focus on iCub's competences and knowledge structure is based on nothing more concrete than pure speculation, an example project selected at random. It is not an attack on the iCub project. Any other "intelligent robot" project is likely to be subject to the same criticisms. We are all instantly impressed by a neat demo, and the underlying reality is anybody's guess.

Moving on from purely verbal stimuli, consider the following two different stimuli of DOCTOR: do they trigger the same stored concept of DOCTOR? Do they trigger the same processes of recognition? Initially, they must be different, mustn't they? Surely the processes involved with recognising a picture are different from those of recognising a word? Would they give the same accuracy and timing results if followed by a NURSE stimulus in either picture or word?

DOCTOR

We could perhaps do a similar experiment with pictures of doctors, nurses and sailors, and perhaps use cartoon type images so that there was scope for misidentification when judgement is rushed. The possibilities for informative probes are endless and the resultant information will not yield sufficient detail to delimit and define a computer-based model of your knowledge organisation.

But I hope that you can see:

1. How it is quite possible to investigate the ways that our minds are structured and operate whilst ignoring the brain.
2. How very difficult it is to get unequivocal information from such experiments.
3. And consequently, how little the scientist can assert with justified confidence about cognitive architecture apart from broad-brush generalities.

Modelling human knowledge organisation quickly becomes mired in a proliferation of links, each associated with a special case or with a general procedure plus exceptions. Everything (whatever those "things" are, e.g., words or concepts) can be associated with every-thing else given an appropriate context. Self-defeating complexity, due in part to unanticipated mutual inconsistencies, always weighs in long before anything like all the experimental evidence has been accommodated in the network model.

Traditional computational technologies have tended to opt for a separation between *data* and *processes*, values of items and the procedures for manipulating those values, respectively. This is closely akin to the distinction between facts and fact-derivation rules. Why make this separation? Primarily to simplify software understanding, and hence aid both initial development and longer-term maintenance.

In the traditional program to compute LIC1, for example, the data items X1, X2, Y1, Y2 and L1 are stored as separate unitary items, and intermediate data values may also be generated and similarly stored — we used the labels O1, O2 and O3 earlier. The procedures used are squaring, square rooting and the greater-than comparison.

Neural computing, as manifest in MLPs, does not make this distinction between data and process. Although, in the equivalent MLP, this distinction is not apparent at this cognitive level, it exists, quite transparently, in terms of lower-level "mechanistic" phenomena — we have a clear distinction between weight-value data and the squashed-sum process. Once again "level" of interpretation is important.

After the initial surge of AI disappointments, the view that lack of knowledge was the missing ingredient gained considerable credence. "After all," AI enthusiasts proclaimed, "how can we reasonably expect an AI system to behave intelligently if it knows almost nothing? Intelligence is founded on knowledge." Knowledge of the world, general and commonsense knowledge such as you and I possess in abundance, was, with uncharacteristic humility, seen as a challenge too far.

It was noted, however, that human experts exhibit some of the peaks of intelligent reasoning based on quite limited and well circumscribed knowledge — expertise implies deep mastery but only of a limited domain. Moreover, the resultant computer experts, being cheaply reproducible, promised considerable societal as well as financial gain.

Highly specialised medical diagnostic expertise, for example, once captured within a computer system could be reproduced for virtually nothing, and distributed widely throughout the (computerised) world.

With this sort of puff the Expert-Systems (ES) or Knowledge-Based Systems (KBS) bubble was quickly inflated by a sudden emergence of KBS start-up companies to augment the purely academic push into this field. Many books have presented the positive spins, the subsequent surprises and the disappointments of this area, and I do not propose to rehash it here. Suffice it to say that many ESs are out and about doing valuable service in the real world (e.g., credit worthiness checking), and many others are not.

Well-defined, independent-discrete-decision expertise (such as credit worthiness checking) proved amenable, but the more diffuse, ill-defined areas (such as medical diagnostics) did not. In these latter domains of expertise, it was believed that, provided human experts in an area were prepared to give their time to explain their expertise, the necessary rules and knowledge could be elicited and used to construct an Expert System — a computer program that replicated the human decision-making capability. It was soon apparent that this was not so. Being an expert and explaining your expertise in terms of precise, explicit and independent decision rules are two entirely different things.

It is curious (but salutary) how one of the early expert systems, MYCIN which was designed to diagnose and recommend treatment for bacterial meningitis, proved to be a dead end (like all the other ambitious medical systems)[8]. Yet it still merits a mention in modern AI literature[9] — not as a cautionary tale, but as a success story! MYCIN did pioneer a strategy for reasoning with uncertainty (one of the points where an approach through simple logic breaks down), but MYCIN's use of "uncertainty factors" solved very little and did not endure either. Even more curious is the supporting "evidence" of MYCIN's success: "In 1979 a team of expert evaluators... found that

MYCIN did as well as or better than any of the physicians." (p. 266) This was written in 2005.

Given that MYCIN and its spin-offs are long dead and buried, and that its creators were not just bored and changed tack after having "solved" the big problem, this result of "success" is odd. The oddity vanishes when one examines the tight constraints on the successful testing with respect to the much broader context within which a useful diagnostic system would need to operate. Bacterial meningitis cases do not arrive neatly labelled; initially they emerge (as perhaps no better than informed hunches) from a complex context of conflicting possibilities and complicating factors. Once more Dreyfus' "intellectual fog"[10] surrounding AI projects is quite sufficient to keep the myth alive and well.

It also became clear that even experts have, and draw on, knowledge way outside their specialist domains in order to make expert decisions. Medical experts, for example, draw on observation and clues from their patient's background or demeanour (perhaps clues that are not consciously registered — a further problem). In many expert areas there was no escape from the need for general knowledge and commonsense knowledge, and lots of both.

One consequence was that in the early '80s an AI researcher, Doug Lenat, secured long-term funding to launch the CYC project to collect together, codify and implement a huge database of facts and rules about the world, a knowledge base of commonsense and general knowledge. This project is atypical of AI projects in that it is still going strong. The majority die quietly, or get reincarnated in a new guise once the researchers have struggled with the original ideas for long enough. Alternatively, one might claim that CYC has been rejigged repeatedly over the intervening decades, but what project does not move with the times? However, the original estimates of the necessity for tens of thousands of pieces of knowledge (facts and rules) have escalated to a need for hundreds of millions. Like the bow wave of a boat, estimates of the amount of knowledge needed to support intelligent reasoning have always surged just ahead of the amount the scientists have managed to build into a computer system.

The original ten-year span for CYC (a remarkably long time in terms of the few-year goals typical of such projects)[11] has come and long gone. CYC has not delivered on any of the grand, original expectations, such as powering an expert system with a sound and relatively complete basis of commonsense and general knowledge. But it has spun off a variety of specialist applications, although many have not yet quite attained their horizon of clear success. Progress has been limited in comparison with original predictions, but steady (dare we say, sluggish) progress there has been.

A major criticism of the CYC project from its outset centred on the difference between "storing" knowledge and using it effectively — "mastery" of that knowledge if you like. "When CYC reaches one hundred million items of commonsense knowledge, perhaps human superiority in the realm of commonsense reasoning won't be so clear." (pages 291–2). This is an uncharacteristically cautious prediction. Yet it still conflates "storage" with "mastery". Between these two states of data there is room for many more AI projects to bridge the undoubted gaps between amassing facts (and further-fact derivation rules), and the sophistication and complexities of using them to support commonsense reasoning.

A major element of the CYC approach to the problems with traditional AI projects designed to capture a goodish chunk of the knowledge that you and I seem to have at our fingertips, or inter-neuronal connections, is to go for more of it. But what is this "knowledge" (commonsense or general)? Is CYC becoming a huge labyrinth whose complexities challenge its designers and deter potential users? What else could it be? What do we know about knowledge and cognitive architecture? Sadly, not a great deal at the detailed level needed to construct the necessary knowledge system.

Indeed, our previous discussion of informationally-encapsulated modularity versus global accessibility of information is primarily a conundrum of knowledge representation and retrieval. Is your brain's knowledge of triangles repeated (perhaps with different specialisations) in a very large number of informationally-encapsulated modules, such as the one we explored in Chapter 2? Or is it global knowledge available to any module that can make use of it?

Consider this little question: what is the area of triangle *ABC*?

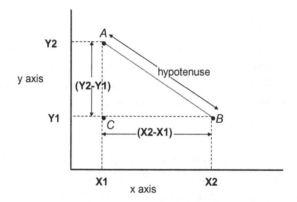

You may be relieved to know that the answer doesn't matter. So what does?

This geometric problem should look familiar. It's just like LIC1 but different. Perhaps you can't recall the formula for calculating the area of a triangle. Perhaps you never knew it? Perhaps you remember that there is such a formula, and you once learned it, but now it's gone. Forgetting is also part and parcel of human knowledge management but not one that I want to explore here.

The point of the above question is to illustrate that you have no problem puzzling over it, whether you can answer it or not. Your knowledge of triangles, however limited, came instantly to mind when I presented you with this little problem. Your knowledge is unlikely to involve an encapsulated module for tackling this problem because you've probably never tackled precisely this problem before. The knowledge of triangles that you now bring to bear on it must be more global knowledge.

But earlier when dealing with the LIC1 problem, we had to insist that the requisite knowledge of triangles was sealed within the module — informationally encapsulated was the expression used to describe the characteristic required. Your knowledge of triangles — is it local to various modules, or global to the full breadth of your reasoning, or both?

The global view accords well with the fact that your brain could effortlessly puzzle over the new geometry problem that involved triangles. It would seem like a general competence is being applied to new information. Does this then become a new encapsulated module for this new knowledge?

If so, we are left with knowledge of triangles as both global knowledge, and knowledge encapsulated repeatedly in a multitude of modules. This smacks of major inefficiency (but could have been evolution's best effort).

Worse, it presents the brain with a major headache (perhaps literally) — with knowledge that's less "fixed" than that of triangles, how is a revision of a general principle about triangles accomplished? The global version can be updated but what about all the local versions sitting in their encapsulated modules? Informationally-encapsulated means that such adjustments to knowledge cannot be introduced, let alone introduced whilst maintaining overall consistency. Apart from positing "down times", such as when sleeping, when otherwise informationally-encapsulated modules are updated, we're in a real fix with informational encapsulation and relevant general knowledge.

About the only thing we can assert with some confidence is that both the storage and mastery of knowledge in the brain are largely mysteries. So why not push on with CYC-type strategies that echo the traditional simplification of separating data and processes. That's still our best guess about the management of the complexity engendered by interacting facts and rules. Whether it bears any relationship to our cognitive architecture and, more importantly, whether it could ever support intelligent reasoning is anybody's guess.

One fundamental problem of knowledge storage that AI researchers have grappled with for years is the slow-down of retrieval as the stored knowledge grows. It does appear to be inescapable that the more knowledge stored, then the longer it will, in general, take to find the chunk that you are looking for — as the chapter-heading quote says, "The more you know, the slower you go". Computer scientists grappling with ever larger databases have expended considerable energy and ingenuity on circumventing this truism.

So-called "hash-coded" search systems[12] in computer science, for example, can maintain very fast access times almost independent of the size of the database. This seeming magic is achieved by: first, using a fixed and known procedure for accessing the information; and second, having a lot more storage space available than is used. Although neither requirement sits well in an AI-system context, there have been innovative suggestions that can offer a relaxation of the requirements in exchange for a degree of retrieval error. Erroneous and inappropriate information retrieval is a feature of human intelligence. Whether it is simply a fault in the human version or somehow intrinsic to the nature of intelligence (an optimal compromise, independent of substrate?) is an interesting issue to which we shall return in the closing chapter.

Clever indexing (storing items with the important retrieval keys attached) can also go a long way towards maintaining retrieval speed as the amount of data grows. But this strategy relies upon knowing what the important retrieval keys are when the data is stored. This need to know in advance blocks the possibility of "creative", or innovative, knowledge retrieval which is, of course, an important component of intelligent reasoning. AI researchers have long fudged this by effectively (if not blatantly) building the knowledge in a way that favours the desired retrieval actions.

A final observation on the problem of knowledge retrieval concerns what we might term negative knowledge, Suppose I ask, "Is the octopus a toy car?" Instantly, you will respond in the negative. How would iCub deal with this request? Presumably neither you nor it happens to have a stored knowledge link to the effect that "octopus is-not-a toy-car" Why? Because if such links were a general feature of knowledge mastery then every node would need thousands of links just to detail all the things that it was not. In the absence of such a link your instant response is a bit of a puzzle because the alternative basis for this answer seems to require that you searched through every path out from the "octopus" node in order to ensure that there was no way in which it could be true. Such extensive searching is hard to square with an instant response without invoking massive parallelism (a possibility that we will consider in some detail in a later chapter).

Google and its competing Internet search engines seem to have gone a long way towards solving the search problem — they quickly and effectively enable us to find relevant information from keys that cannot all have been foreseen. Is this problem almost solved? Not really. Despite the amazing knowledge discovery and management that so many of us treat as routine today, it does require an intelligent filter (i.e., you) to extract the nuggets from the heap of ore typically produced. Current Internet search is amazing but only because it combines the computer's muscle (in conjunction with predefined "keys" or "tags") with the user's brain. The automated searching is syntax, i.e., keyword, based. The meaning that identifies the relevant gems is still intelligence based. Meaning-based information retrieval, semantic information retrieval, has long been a goal of computer scientists, but it seems that we might need AI first, which takes us full circle.[13]

Thus we had the spate of intelligent "rediscovery", rather than "new discovery", systems of some years back[14]. If you know the "discovery"

234

that you're aiming for then it is not too demanding to construct a system with elaborate and interesting methods of making the "rediscoveries". But building a system to really discover new and interesting things is something altogether different, although the hope must have been that success with "rediscovery" would give useful clues about "discovery". Apparently that did not happen, if the subsequent dearth of discovery systems is any guide.

A claim for one significant new theorem discovered by such a system is often declared as some sort of "proof" that AI is forging ahead into the deepest realms of human intelligence. Quite apart from the peculiarity of just one, or even a few, such proofs (if such an AI system really existed we might expect a deluge of new and significant proofs from such an "expert mathematician" working around the clock, seven days a week), Dreyfus dispelled the mists from around this claim in 1972, and revealed it to be a long-known and rather simple proof[15].

But the "intellectual fog" movement rolls on. Always dependable, it has, for decades, trotted out the two "great discovery successes" of AI systems in order to prove that AI was realised years ago. The two proofs are: the discovery of a major molybdenum deposit on Mount Tolman in the USA by the PROSPECTOR system[16], and the discovery of a proof of "a long-standing conjecture in Algebra... by an AI system... Human mathematicians called the proof 'creative'." (p. 283, but no reference to the source, probably the automatic creative act debunked by Dreyfus above). In both cases there is the issue of how the system worked and how the commentators say it worked — these are never quite the same, and sometimes a good deal of wishful thinking combined with the necessity to omit detail means that significant distortion arises[17]. But the ultimate test, the one which cuts through all the scope for bluster and misrepresentation, is, of course, how many further creative or intelligent insights have these programs produced? The answer is none, and the systems have long been abandoned.

Together with the MYCIN story, these one-shot successes are emblematic of a recurring element in the intellectual fog of AI.

A "star" AI system is reported as achieving one significant-sounding success, a breakthrough; then it vanishes from the active scene having claimed a permanent place in the AI hall of fame as key evidence for the success of AI. We have already noted Deep Blue's one-time defeat of Garry Kasparov. IBM's Watson, a game–playing system, is set to be the latest high-profile addition to the roll call of honour (as we see later).

If any of these systems achieved its success as a result of the applications of significant underlying principles of intelligence, we might expect these principles to be subsequently re-applied (at least) equally successfully. Yet no further breakthrough successes are recorded. What are we to make of this variety of one-shot success behaviour? The cautionary aphorism concerning the presumption of summer based on single swallow sightings springs to mind.

Progress in a science is based upon the discovery of general principles, and a general principle by its very nature should support further specific successes. When this does not happen, what is to be concluded? The first-success phenomenon is due to specific engineering of the system, and not general principles at all, or it is a fundamentally fortuitous outcome. Neither basis can form the stuff of science.

By its fruits you shall know it. This maxim (number 8), which is typically associated more with religious activities than scientific ones, is, however, apt in this instance. The various one-shot AI breakthrough successes get trotted out in virtually every new presentation of the field with a glaring absence of follow-up successes. There are no fruits from the supposed breakthroughs, and that's how you know their real worth.

It beats me why mention of an isolated success of a decades-old AI system can be presented as positive evidence for the health and vigour of the science. To me, it emphasises precisely the opposite — the time-worn, one-shot nature of the evidence is testimony to the absence of scientific progress.

One is reminded of the time of the blossoming of projects to teach chimps and gorillas human language, usually in the medium of signing. Within the subsequent welter of disappointing results, one gorilla who had taken a liking to watermelon but did not know the name of this fruit asked for "drink fruit" by combining two symbols that she already knew. This creative act caused quite a splash. It was a first. Sadly, it was also a last (as far as one can judge from the absence of follow-up publications). One creative (or fortuitous) insight does not confirm a theory, except in AI.

More innovatively, we can now construct holistic representations of knowledge integrated inextricably with the processing rules (e.g., in an MLP). As we have seen, this buys us almost nothing but problems because of the encapsulated black-box nature of these representations. Hardly a promising alternative to the piecemeal approaches to knowledge representation that have dominated AI, unless of course we can discover some way to share the knowledge across huge numbers of such modules while at the same time maintaining encapsulated modularity... on the face of it, a very tall order.

Endnotes

[1] Larry Page, the founder of Google, is quoted in *Googled* (Penguin, 2009) by Ken Auletta, p. 322.

[2] *The Singularity is Near* (Duckworth, 2005) by Ray Kurzweil, p. 261.

[3] A complaint voiced by frustrated AI researchers as they witness their slick little system get bogged down by all the bells and whistles (and extra knowledge) gradually added (or automatically learned) to elevate their initial system towards its hoped-for levels of intelligent performance. I used it as a subsection heading in *A New Guide to AI* (Ablex: NY, 1991) p. 443.

[4] In 1950 K. Lashley published "In Search of the Engram", in *Psychological Mechanisms in Animal Behavior*, Symposia of the Society for Experimental Biology, no. 4, pages 454–483 (Academic Press). Lashley performed a gruesome series of experiments that involved training rats to transverse a maze (therefore they had learned how to do this traversal), then removing a chunk of the rat's brain (under anaesthetic), and then retesting in the maze (after recovery from the surgery) in an effort to discover whereabouts in the rat's brain the learned memory was stored. He

never found it. Roughly summarised, the result was that the rats continued to struggle through the maze with a deficit that correlated with the amount of brain removed. The memory, the engram, appeared to be dispersed throughout rather than localised in any one part.

[5] For those who want more see, for example, Edward E. Smith & Stephen M. Kosslyn's *Cognitive Psychology: Mind and Brain* (Pearson/Prentice Hall, 2007). Almost any Cognitive Psychology textbook that is, say, post 2005 will provide very similar surveys.

[6] We must neglect all the deep complexities of "true knowledge" and "knowing" here, and so, for example, dismiss the possibility that the "purple car" is in fact, say, green in colour. Knowledge management in the absence of all peculiar possibilities is still way beyond the current state of the art, so we do not need to delve deeply into the intricacies of "knowledge" in order to make the necessary points.

[7] The instructor speaks into his microphone which suggests that iCub "knows" his personal pronunciation patterns. It is plain to see that we all speak surprisingly differently when speech signals are examined, and the best of the current speech-recognition systems are trained to be optimal on a particular person's, or few persons', speech idiosyncrasies. Curiously, the weakness of intelligence is in the opposite direction — we are not so good at distinguishing between the speech patterns of different people. In this bright new millennium apparently there are practical speech recognition systems that do not need to be pre-trained to a user's voice patterns. The latest (as of 17/7/2012) Apple iPhone boasts, "Siri on iPhone 4S lets you use your voice to send messages, schedule meetings, make phone calls and more. Ask Siri to do things just by talking the way you talk. Siri understands what you say, knows what you mean and even talks back. Siri is so easy to use and does so much, you'll keep finding more and more ways to use it." See http://www.apple.com/uk/iphone/features/#siri. Siri, you'll not be too surprised to learn, can also be very frustrating at times, but it does seem to be a significant step forward in voice recognition technology and language understanding.

[8] I do not think that "dead end" is too harsh a judgement; MYCIN did evolve into EMYCIN, a tool for developing expert systems, and a wide variety of similarly defunct expert systems were explored. It is probably true that some IT systems out and about and doing a good job can claim a tenuous link back to MYCIN, but all are a far cry from the original goals of MYCIN.

[9] Kurzweil repeats MYCIN's blind-test success over the expert human diagnosticians (p. 266) without saying why MYCIN was then abandoned.

[10] From Hubert Dreyfus' *What Computers Can't Do* (Harper & Row: 1972) an early extensive critique of AI that was considered in some detail in Chapter 7.

[11] Addressing the bluster of hopeware science in AI in the early '80s AI researchers used to say jokingly that Doug Lenat, CYC's main man, deserved to be immortalised by enshrining his name as the measure of "bogusity", a bogusity index calibrated in

Lenats. But for practical use, to measure most everyday bogus claims, the scale needed to be calibrated in *micro-Lenats*. Such coffee-time talk at AI conferences was perhaps motivated by jealousy of Lenat's coup of a ten-year funded research project, CYC, when most others were overjoyed to get funding for three years. Maybe now, in this bright new millennium, when Lenat is still in business with CYC and most of the others dead in the water, this joke needs re-evaluating (one way or another).

[12] Hash coding processes the item to be stored and generates from it an address (by "hashing" the item) at which to store the item. So a knowledge of the crucial keys of an item (and "hash" procedure used) permits a retrieval system to compute, and inspect the location where this item will be with no searching at all. Every conceivable item that might need to be stored must "hash" to a unique address (otherwise there will be "collisions" when two items "hash" to the same address). Hence, the database will always have empty locations available for future items. How much extra storage is needed depends on the nature of the items and the hash procedure used, and hence their hashed addresses, as well as the expectations with respect to future additions. Many embellishments of basic hash coding have been proposed to circumvent the various limitations. In practice, the possibility of "collisions" must be accommodated, and one "off the wall" (with respect to standard computing, but may be not for an AI system) proposal was for "allowable error" to be introduced by not resolving collisions and thus saving both space and retrieval times (see "Space/time tradeoffs in hash coding with allowable error" by B. H. Bloom, published in *Communications of the ACM*, vol. 13, no. 7, July, 1970).

[13] Updating the quest for semantic searching, i.e., meaning based rather than structural key based, we find repeated mention of the "semantic web" but no search engine based on the "meaning" of your query rather than its "structure", i.e., the precise words and phrases you use, is yet on the horizon despite a few claims for breakthroughs and decades of promises for semantic information retrieval.

[14] The golden age of "redicovery" spawned, for example, AM and EURISKO ("Why AM and EURISKO appear to work" by D. B. Lenat & J. S. Brown in AAAI-83 Conference Proceedings, Washington, DC, pages 236–240, 1983) and BACON ("Computer modelling of scientific and mathematical discovery processes" by H. A. Simon in *Bulletins of the American Mathematical Society* vol. 11, pages 247–262, 1984).

[15] In his extensive exposé of AI claims (he took the trouble to chase them down to their origins — all fully documented in his book, Endnote 10), Hubert Dreyfus states that "there is a striking disparity between Ashby's excitement and the antiquity and simplicity of the proof" p. xxx.

[16] In September 1983 anchor man Dan Rather on CBS Evening News broke the astonishing news of computers finally becoming intelligent. The Dreyfus brothers, Hubert and Stuart, looked into the claim and found, of course, that the reality was much more mundane — support for their analysis lies in the fact that during the intervening decades the PROSPECTOR system has found nothing worth reporting.

The full details are available in the book *Mind Over Machine* by Hubert and Stuart Dreyfus (The Free Press: NY, 1986), pages 115–116.
[17] Although it is far from amusing to pore over and comb through somebody else's complex program, a number of AI systems have been subjected to such scrutiny. In fact, frustration with published accounts of a few "intelligent" results generated by large and complex programs led to an element of AI methodology known as "rational reconstruction" — reproduction of the essence of an AI-system's salient behaviour with another program constructed from the published descriptions. Doug Lenat's AM system of the mid-1970s is one such system that was examined. AM "discovers" concepts and conjectures in elementary mathematics. The conclusion of one such exercise (G. D. Ritchie & F. K. Hanna, "AM: A case study in AI methodology", pages 247–265 in *The Foundations of AI: a Sourcebook* edited by D. Partridge & Y. Wilks, Cambridge University Press, 1990) was that Lenat's "written accounts give a misleading view of how the program worked".

❧ Chapter 10 ☙

Learning Machines — Climbing Lost and Blind

"Machine learning is a complex problem composed of very many, very complex subproblems."

Geoff Hinton, 1990s[1]

"We... already are building 'machines' that have powers far greater than the sum of their parts by combining the self-organizing principles of the natural world with the accelerating powers of our human-initiated technology. It will be a formidable combination."

Ray Kurzweil, 2005[2]

As far as the enquiring mind is concerned the mystery and wonder conjured up by the revelation that we now have machines that learn is nearly always enough to put paid to scepticism about intelligent machines. It's an eye-opener that clinches belief in AI as a current reality.

In 2011 Kevin Warwick, a professor of Cybernetics, used just such a disclosure to laughingly put right the amateur misconception [*sic*] of his interviewer, Michael Buerk[3]. Warwick stated that "a lot of machines actually learn and adapt" which was enough to quash any further discussion of the dissimilarity in learning and understanding that exists between people and machines.

The unavoidable, but erroneous, implication is that AI is about to be realised, here and now. Warwick dismissed objections to this point from philosophers and computer scientists as "not technically based". Well, here comes a technically based one, but couched in terms of mountaineering. This is a field of human endeavour that every reader can appreciate (in terms of the generalities we use), even if they view it as no more than a minor lunacy that infects a small subset of humanity.

Machine learning, which is the general term for all attempts to develop learning strategies for computers, is viewed pretty much exclusively as a mountaineering exercise. How can this be? In an earlier chapter we saw learning in neural networks portrayed as climbing up a hill. Now we can broaden this special case to become the general basis for all machine learning.

Learning is a process of improving some aspect of system behaviour — we may want a faster system, a less error-prone system, a system that can deal with changes in its environment, and so on. We would like a computer system that can, as a result of experience, get "better" by itself (or at least get no worse as new challenges arise in its operating environment). There are countless reasonable interpretations of the crucial term "better" and it must be objectively defined for each learning procedure developed.

So where do the pitons, the ice fields, the overhangs and the death zone of mountaineering fit in? They don't, but if we trivialise this noble exercise as one of choosing ways to climb up mountainsides with the better climbers reaching the highest peaks, we have machine learning in a nutshell. The best learning systems climb up to the highest peaks, and the higher the peak, the better the system has learned.

It is overwhelmingly height that determines the quality of a machine that has learned; how it got there is unlikely to be important. So learning machines have no truck with choosing difficult routes just because they are there. The North Face of the Eiger, for example, would have been ignored. The existence of difficult routes upwards is also a challenge for learning machines, but the challenge is to find an easy way up around the difficulties. For a computer system, getting to the top, and to the highest top, is all there is to it.

Let's suppose we are aiming for a machine that will learn to become more accurate. A first requirement is to define what we mean by "more accurate". "Accuracy" might, for example, be defined in terms of the number of wrong answers produced within a 100 test cases. Then a "more accurate" machine will get fewer of the 100 test cases wrong than a "less accurate" version. Under this definition, a "better" system delivers correct answers to more test cases.

Within this framework we can give substance to the mountaineering metaphor. The very bottom of the mountain is a version of the system (using whatever technology we choose) that gets all 100 test cases wrong. There can be no worse system, can there? Let's call this the 0-system. On the mountain peak sits the very best version, the one that gets all 100 test cases correct — the 100-system.

Here's our mountainside:

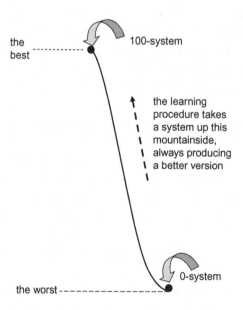

What is this steep curve? Recall from the chapter on self-adaptive, i.e., learning, systems, or even better just think about it all afresh — a learning system must change itself. So during a gradual process of learning we get a sequence of different versions of the same basic system with each new version performing "better" than its predecessor (so each new version sits higher up on the slope). Learning is thus a process of moving steadily up the mountainside. It is mountain climbing.

For example, in the case of an MLP neural-network system, which we explored in an earlier chapter, the learning procedure is one of altering the link weights. So the sequence of versions will be the same basic network (same units, same links) but with different weight values on the links.

A proper learning procedure (as opposed to say random link-weight changes) produces a sequence of better and better versions of the neural network system. Such a learning procedure can be viewed as one way of moving the basic system upwards on the illustrated mountainside. Every position on this mountainside represents a different version of the basic system, a different set of link weights that gives a particular accuracy when the MLP system is tested.

Within this general framework, the final challenge is to devise a learning procedure for the particular system technology being used. For the MLP system we have the Backprop Algorithm, a learning procedure that modifies link weights with guarantees that successive versions will never get "worse". For a learning database system we might have a learning procedure that automatically adds new facts, deletes out-of-date ones, and perhaps rearranges the whole lot to optimise its performance on an ever-increasing amount of knowledge.

And so on. There are many programming technologies and even more possibilities for learning procedures to automatically change their behaviour. But learning procedures that always deliver better versions are in short supply. We might call these "good" or "proper" learning procedures[4]. I used "well-founded" to describe this property of the MLP learning.

Given the challenge of devising a good learning procedure, machine learning would seem to be plain sailing, or straightforward hill-climbing (the tradition is to talk of climbing hills rather than mountains). But it is not. Why not?

For a start, the learning landscape is never one simple peak. It tends to be a complex mountain range of peaks (of various heights), valleys, ravines, plateaux, etc. Here's a relatively simple mountain range, but one that's complex enough to reveal some of the principal difficulties faced by learning machines.

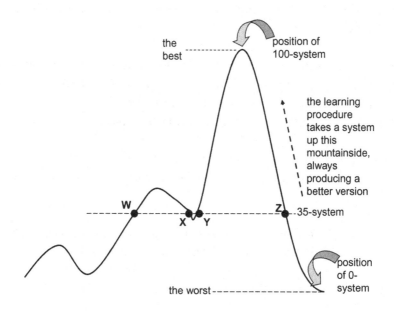

In this more elaborate mountain view, the simple up-slope has become part of a peak, and two further peaks have been added. In addition, at the height of an accuracy of 35 correct test cases of our 100 (the level of the 35-systems), I've added a broken line. You will see that it cuts through our landscape in four places. This means that there are four different versions of our system that each gets only 35 of the 100 test cases correct. These four different versions of our system are labelled **W, X, Y** and **Z**.

Now suppose we have either version **Y** or version **Z**. The learning procedure will push the chosen system up its respective slope until it sits on the highest peak. In either case it will have learned to become the best possible system, the 100-system.

Suppose we start instead with version **X** of our system. It will also learn and so climb up its slope to reach its local peak which is a good deal lower then the highest one. This peak may be a 45-system, but it is a peak, and our system sitting atop this summit can learn no more because there is nowhere upwards to go (such is the nature of peaks). What does this mean?

It means that our initial system **X** did begin to learn, and improved its performance from 35 to 45 correct test cases. But then it stopped learning. It became stuck on this sub-optimal peak. A proliferation of such sub-optimal peaks often plagues machine learning exercises.

"Why is this a problem?" you might ask. "Just start with systems on the sides of the biggest peaks and avoid versions such as **W** and **X** in our illustration."

This is the obvious solution, but I have yet to mention one further difficulty that bedevils machine learning — although the learning landscape may be the fairly simple one illustrated, neither we (the scientists), nor our learning systems, can ever know this.

The mountain range that the system is climbing in is unexplored and permanently obscured by clouds. Our learning system must climb as if blind in an unknown landscape. Here's something a good deal closer to the reality of machine learning.

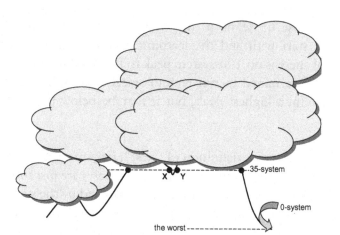

Within this harsh reality of an unknown and unseeable landscape, getting to the top of the highest peak has suddenly become more difficult, much more difficult. Now the choice between version **X** or version **Y** (both 35-systems) as the best one to learn from is no longer straightforward. A sensible and commonly used strategy is to choose the version on the steepest path in the hope that this will lead to the highest peak. So, when the choice is between versions **X** and **Y**, the learning system will choose version **Y** to learn from in the above illustration. With our special omniscience, we know that this is the better choice, but that's pure luck.

It is as if you, the mountaineer, have been deposited in the middle of a black night on an unknown mountainside with the task of climbing to the highest peak. What would you do with no GPS and no smart phone signal? A sensible strategy (apart from giving up) would be to feel around your immediate vicinity and start up the steepest path. This is what learning machines generally tend to do.

The learning system does not know where the 100-system peak is. Indeed, it may not know whether any 100-system peaks exist in this particular mountain range. It may be that the "best" system, however we may choose to define it, may not be achievable with the basic system we start with and the learning procedure being used. This means that there is no 100-system peak to be found within the mountain range determined by our technological choices. There will be a "best" system, a highest peak, but it may be below the 100-system optimum.

In terms of our earlier definition of "best", i.e., no wrong answers for 100 test cases, it may be that a couple of test cases are just too difficult for any version of our basic system (as determined by the system technology and the learning procedure we are using). No version can compute them correctly. In which case, the highest peak attainable may be a 98-system peak.

Given this uncertainty associated with sub-optimal peaks (is a 98-system peak the highest?), the learning machine cannot know whether it might do better to start again with a different version, or whether it will be wasting its time. As we shall see in the following chapter, the so-called "evolutionary systems" directly tackle this difficulty.

Once the full picture is revealed, you might want to concede that the human mountaineer has it easy. With detailed maps and similarly detailed analyses (or even reports from previous climbers) of possible routes upwards, no wonder he or she has to choose hard routes in order to spice up the climbing.

It is similarly easy to see why learning machines do not always reach the highest peaks. Indeed, given these obstacles, it is surprising that high peaks are ever summited, and very often they are not.

The take-home message here is that, in general, a learning system does not know if the peak of achievement it seeks exists, or if another version might be able to climb higher. Climbing blind in an unexplored mountain range always means that failure to reach the highest peak may be due to climbing in the wrong part of the range, wrong choice of upwards route, or the absence of such a peak within this mountain range. All in all, this is not a very satisfactory state of affairs.

So how do you and I crack this problem? We learn all the time, and constantly do better in all sorts of ways on all sorts of tasks. Maybe it's all achieved by simply reinforcing inter-neuron connections every time they are used. Maybe this principle underlies all that we learn. But even if true, exposure of this underlying principle of how our brains learn contributes virtually nothing to our knowledge of how to design and build an intelligent learning system.

While there is at least a grain or two of truth in this glowing characterisation of human learning abilities, it is not that simple, is it? We often fail to learn useful information, we sometimes "learn" things that degrade rather than improve our performance, and seldom can we be sure that we've learned in the best way possible. Human learning is not perfect... far from it.

Nevertheless, as things stand, the learning integral to basic human intelligence is a wonder of multi-dimensional (i.e., we seldom work to just one clear measure of "better") and constant improvement. There are machines that learn but they barely begin to scratch the surface of the complex learning that you do (much of it quite effortlessly), and do constantly on a minute-by-minute basis.

With learning machines struggling along in a pitch black and unknown, low-dimensionality landscape, they can only gaze in awe (if they had the wherewithal to do such things) at the masterful performance of the human toddler. This human is learning both the intricate complexities of its native language and the multi-dimensional ways of its world simultaneously.

Nevertheless, science has made, and continues to make, slow progress with learning machines. In the next chapter we will examine a couple of the best learning technologies, and see exactly how they do their mountaineering in order to try to reach the highest peaks.

Endnotes

[1] A loosely remembered quotation from a talk by eminent AI researcher, Geoff Hinton.

[2] Ray Kurzweil's *The Singularity is Near* (Duckworth, 2005), p. 483.

[3] The UK's BBC Radio 4 *The Choice* on Tuesday June 14th 2011 with Michael Buerk.

[4] As usual, the reality of machine learning may need to be more complex — human learning does not always immediately deliver a better system. For example, when learning to perfect a motor skill, say, serving in tennis, it is not unusual for the subject's serve to first deteriorate before eventually getting better. Once more, we have an example of human learning being a good-enough compromise (between getting better and not immediately getting better) rather than being an optimal solution (i.e., an inexorable process of getting better and better in all respects).

☙ Chapter 11 ❧

Hot Technologies — the Doomed and the Dubious

"The task of creating human-level intelligence in a non-biological entity will involve... a learning, chaotic, self-organizing system, one that is ultimately biologically inspired."

Ray Kurzweil, 2005[1]

"Set up a gene pool of possible software and let the successful program evolve out of the soup." [author's emphasis]

Steven Johnson, 2001[2]

"The standard current alternative to Turing architecture, namely connectionist networks [i.e. neural computing], is simply hopeless. Here, as so often elsewhere, networks contrive to make the worst of both worlds... It must be the sheer magnitude of their incompetence that makes them so popular."

Jerry Fodor, 2000[3]

With the highs and lows of mountaineering machines firmly in mind, it is time to see exactly how the "new", biologically-inspired learning technologies have been configured to deal with the problems, such as sub-optimal peaks. More generally, what can they contribute to our understanding of this core component of intelligence? What are these hot prospects for opening up new and more productive possibilities for AI?

Neural computing is an obvious candidate, if for no other reason than the fact that the only intelligence we know of runs on neurons. The superficial match is incontestable, although it is also pretty much devoid of real substance. However, we've already uncovered much of the uninspiring reality of neural computing. Now we can see it through to its unpromising current status as a basis for a scientific breakthrough in our understanding of intelligence.

A second technology that attracts much hopeful attention is the one based on so-called Genetic Algorithms (GA). Again the nomenclature does it no favours. With its suggestion of a close relationship with that other hot topic, genomics, and the hint (if not more) of an evolutionary basis, it is no small wonder that those who understand little of how GAs work are wildly over-optimistic about them. With justified wonder, we find some technologists similarly smitten[4]. So GA technology (aka "evolutionary" computing) also needs to be examined in order to appreciate its scope and expose its limitations.

Jumbled in with these seductive technologies we encounter "chaotic", "emergent" and "holistic" possibilities as well as the anticipated powers of "self-organisation". The latter two have had their own chapters, and so will not be revisited except where unavoidable. Perhaps all that the first two need is tying down in terms of less emotive words. So, let's start there.

"Emergent" properties, whatever they are, usually provide another lifeline of hope when it's unclear how to get what you want from a system or how to explain what you observe. So what's "emergent" in this context?

Steven Johnson's popular 2001 book *Emergence*[5] is of relevance to our enquiry because he addresses software systems as one of his major areas of interest. In the book we find "emergence" defined as: "The movement from low-level rules to higher-level sophistication" (p. 18). Accepting the haziness of "higher-level sophistication", we might take it that Johnson would see intelligence, manifest as cognitive-level behaviours (the "higher-level sophistication"?), as an emergent property of the low-level rules such as those we exposed as governing neuronal information processing.

So what is the gain here? Has the revelation of "emergence" done anything except rename the current mystery of the basis for this "movement"? Is "emergent" just a place holder for "don't yet understand"?

For example, before molecular energy was understood, "heat" was an emergent property of certain substances and mixtures. In the modern era, heat is entirely explainable in terms of energy levels and transfers at the molecular level.

Maybe toast is an "emergent" phenomenon with respect to the Cosmic Toaster, but not with respect to the electric one? Or does use of the epithet "emergent" also carry the implication that the mystery is fundamentally unfathomable for some unspecified reason? I don't know.

Johnson does mention "laws of emergence" on several occasions, but puzzlingly omits any statement of what these laws might be. We are left in the dark. Nevertheless, he is clearly impressed by the emergent murkiness he perceives in "bottom-up, evolved" programs "where a kind of digital Darwinism [evolutionary/genetic algorithms?] leads to a simulated intelligence, capable of open-ended learning" (p. 173), and then presumably the sky's the limit — soon, I expect.

However, as we'll see below, we really do know (and have done so for decades) how and what behaviours will indeed emerge, and it's not intelligence, simulated or otherwise.

For Ray Kurzweil, "emergence" is also clearly a good thing, although not defined explicitly. In response to the Singularity critic who suggests that intelligent life is based upon system properties that computer systems cannot exhibit, Kurzweil says it is over-hasty to exclude the possibility of physical systems exhibiting "emergent... vital characteristics of organisms such as self-replication... and the holistic order of biological design". (p. 483). This critic, he claims, "is ignoring here the ability of complex processes to exhibit emergent properties that go beyond 'its parts in isolation'... nothing restricts patterns and their emergent properties to natural systems" (p. 480).

Unusually, perhaps uniquely, I find myself squarely behind Kurzweil here. Every computer program exhibits properties that go beyond its statements in isolation. That's the whole point of putting bits and pieces of a programming language (such as statements, keywords, named variables) together to produce a coherent whole — a working program. Programming can be cast as the process of piecing together isolated parts in order to reap the benefit of the emergent properties of the whole. When we add in our Maxim 9 — **no significant computer program is completely understood** — we also pick up the unfathomability aspect (but stripped of its mystery).

Taking this viewpoint, "emergence" with respect to software systems is the norm rather than some inexplicable exception. The reality is that "IT-system behaviour is completely and precisely defined, but effectively unfathomable in its entirety"[6], and in that (not very enlightening) sense, *all* software system output is emergent. This also fits in nicely with Johnson's definition, although he seems to want to reserve it for the more mysterious (to him) computational technologies.

Positing "emergent" properties of ill-understood software technologies gets us nowhere in the quest for AI. It is no more than an admission of ignorance, which, although sometimes commendably honest, cannot be a cause for celebration on the assumption that therefore AI must be just around the corner. Out of ignorance compounded by obfuscation, AI is unlikely to emerge.

"Chaotic" computing is indeed a bit mysterious to a traditional computer scientist, but it hardly sounds like something to be welcomed in a software system. In its everyday meaning it is a quality that the software engineer is daily fighting tooth and nail against.

There is, however, a class of neural computing networks, notably the so-called Boltzmann machines and their kith and kin, in which an initial "chaotic" state is systematically "cooled down" (called "simulated annealing") to result in a stable pattern of activity such that an answer can be "read out"[7]. However, this type of neural computing, which is totally different from the MLP type, is difficult and somewhat hit-and-miss to set up, and consequently not widely used. Every Boltzmann-machine network is a problem-specific creation that requires a technically competent and dedicated creator with time on her hands to set it up.

Because the behaviour of this class of neural network is governed by a probabilistic equation (as well as differing approximations to the infinitely-slow annealing that the theory specifies as necessary to give the very lowest energy solutions, these are the best ones), they exhibit the property of non-determinism. This is non-determinism in the sense that different patterns of stable activity (and hence different solutions) emerge on different annealing runs of much the same initial set-up. There is a certain amount of interest to be extracted from this non-mysterious variety of solutions available[8], but not such as to merit the glee with which this non-determinism is sometimes greeted by ill-informed AI enthusiasts.

To these incautious positivists it is another computational property that echoes a fundamental of human intelligence, and it is thus a good thing, and, moreover, one that points the way to anticipated breakthroughs. To the technologist at the code face, it is no more than an invitation to experiment with annealing strategies, and to repeat problem-solving runs in an effort to find the most efficient strategies that consistently deliver the best results.

The Boltzmann machine is representative of an interesting class of computational technologies, one that probably hides significant developments that are yet to be discovered[9], but as a possible aid to AI it is hopeless — a Boltzmann-machine network (just like an MLP but much more tricky to configure) computes just one fixed process. There is no known sense in which this technology can significantly contribute to a flexible sequence of reasoning[10], not even if we return to the holism escape route.

The brain (to put it in CTM language) is capable of computing many different processes, so it can't be a single Boltzmann-machine-like system. Alternatively, if the brain is viewed as a vast collection of these one-shot modules, what manages this collection of independent computers (quite apart from any vaguely plausible mechanism for configuring all these individual "machines" in the first place)?

We are right back to the local-modular versus global impasse, except that now we've made it worse. Now it is further complicated by the need to individually configure each Boltzmann-machine module (unlike the MLP modules which are easily set up and trained — albeit implausibly as human learning — to learn specific processes). It is also necessary to manage the activation and monitoring of the necessary simulated annealing regimes (whereas a trained MLP is just activated by input signals, and computes an output signal).

I do think that advocates of the merits of chaotic computing as well as of the positive expectations for emergent properties have an obligation to be more precise. As a scientist or engineer you cannot get away with just waving the words around to cloud the argument and so claim a "one bound" escape from the difficulties that every precise analysis uncovers. Mystery does have its attractions, as Agatha Christie fans will testify. But the reader, just like the scientific enquirer, has a right to feel cheated if subsequent attention doesn't lead to a coherent story and clues (at least) as to "who dunnit", or in our case, "how it is done".

Having disposed of the more emotive terms, we are free to focus on the "new", biologically-inspired technologies — all of which are really quite long in the tooth on the timescale of electronic computation. First let's sort out a technical term that we've made some use of but now need to pin down more precisely. It is used in two rather different senses, but senses that can be reconciled. I refer to "functions".

The everyday use (for those readers whose everyday lives are sufficiently peculiar to use the word "function" regularly) in the context of cognition refers to the various components of intelligent behaviour: seeing, speaking, etc. And to further complicate this fuzzy definition, each of these major "functions" will be divisible into component "sub-functions". So the "seeing" function might be composed of a "retinal image processing" function, followed by an "image-element composition" function, etc. Then "seeing", as a functional component of intelligent behaviour, takes retinal images as input and produces an interpretation in terms of world objects as output. The pattern of light falling on a retina becomes transformed (we know not how) into, say, a mental picture of a horse standing under a tree in a field. The "retinal-image processing" sub-function accepts the same input but transforms it into "image elements", say, lines, curves, and shapes. Further sub-functions take these lines, curves and shapes and output, say, named objects like horse and tree, etc.

This is all dreamed up. The point being that some breakdown in terms of separate functions is a sensible and commonplace way to go about trying to comprehend and rationalise how the intricacies of cognition (or indeed any complex process) come about. This is so even though we are well aware that the assumed functions and sub-functions of cognition, for example, are probably not strictly functions in the formal sense. They are not informationally-encapsulated because they both leak and absorb information by means other than their designated inputs and outputs.

The essence of a function is that it transforms inputs into outputs without any intermediate inputs or outputs. Our original brain module implemented a function, and the goal of the reverse engineering is to determine what that function is at the cognitive level.

"Function" is also a well-defined notion in mathematics, and, it is pleasing to report, it is essentially the same notion as introduced above. So we are free to discuss functions with precision but without any off-putting mathematical foreplay.

Formally, a function can be defined by either stating its behaviour (in well-defined terms) or by listing its input-output pairs. Thus, the function "doubling the addition of any two integers from 1 to 3" — we'll call it ADD3TWICE — may be defined as:

ADD3TWICE is $(m + n) \times 2$, where m and n can each have a value of 1, 2 or 3.

This definition relies on the accepted definition of the arithmetical operators "+" and "×" as well as the meaning of the parentheses which specify that m is added to n before the multiplication by 2.

Alternatively, we can say that ADD3TWICE is defined by the following set of input-output pairs — inputs m and n generate output p, written $m, n \rightarrow p$ below

$1,1 \rightarrow 4$	$1,2 \rightarrow 6$	$2,2 \rightarrow 8$	$1,3 \rightarrow 8$	$2,3 \rightarrow 10$
$3,3 \rightarrow 12$	$2,1 \rightarrow 6$	$3,1 \rightarrow 8$	$3,2 \rightarrow 10$	

These nine input-output pairs also *define* the ADD3TWICE function because there are only nine possible pairs of input values. This tabulation of the ADD3TWICE function is sometimes called a "look-up" table definition.

If, as I claim, we now have two definitions of the ADD3TWICE function, then we should be able to use either definition to determine if any two integers are "in" this function, and if so, what is the corresponding output value. Let's just do this.

Is the pair (1, 4) in function ADD3TWICE? No. It is excluded by our descriptive definition because one of the proposed inputs, the "4", contravenes the stated restriction that both m and n must be 1, 2 or 3. Alternatively, our look-up table definition excludes this pair of numbers simply because when we check all nine entries, the input pair (1, 4) is not found.

What about the input pair (3, 2)? This time the answer is "yes". Moreover, our definition tells us what the corresponding output is: it is 10. How do we know these things?

If we use the descriptive definition, we find that $m = 3$ and $n = 2$ constitutes a valid input pair. Then $(m + n) = 5$ by the rules of addition, $5 \times 2 = 10$ by the rules of multiplication. Hence the input pair (3, 2) corresponds to (or produces, or generates, or maps to) the output 10 in the function ADD3TWICE.

What about if we use the look-up table definition? We search through the table entries looking for the input pair (3, 2), and we find them as the last entry. As well as this, the table entry tells us that the corresponding output is 10.

For ADD3TWICE I devised the handy limitation on the input values by allowing only three possible values for m and n. If we remove this limitation and define a new function, which we'll call ADDTWICE, as "doubling the addition of *any* two integers", then I can still define this function as:

ADDTWICE is $(m + n) \times 2$, where m and n can each have any integer value.

But I can no longer *define* it by listing its input-output pairs. Why? Because there are an infinite number of them. I could give a *sample*, though:

$-3,3 \to 0$	$-10, -5 \to -30$	$0,4 \to 8$	$1,2 \to 6$	$11,11 \to 44$
$17,23 \to 80$	$21,3 \to 48$	$44,66 \to 200$	$99, 101 \to 400$	$167,13 \to 360$

These 10 input-output pairs are part of the definition of the function ADDTWICE. They are a sample from all the possible pairs that constitute the ADDTWICE function. A little surprisingly perhaps, they are also a sample from the definition of any number of other functions. Here are a few of these other functions:

1. $(m + n) \times 2$, where m and n can have any integer value greater than −11
2. $(m + n) \times 2$, where m can have any integer value less than 168
3. $(m + n) \times 2$, where $(m + n)$ cannot have the value 33
4. $(m + n) \times 2$, where $(m + n)$ can only have the values 3, 24, 0, 100, 40, −15, 200, 22 or 4
5. $(m + n) \times 2$, where neither m nor n can have the value 403
6. $(m + n) \times 2$, where $(m + n)$ cannot have the value 25

So our sample of input-output pairs for the ADDTWICE function is also a sample from the functions defined in 1 to 6, above, as well as a sample from countless other functions.

We are now fully equipped to complete our survey of MLP technology.

As described in Chapter 6, we obtained an MLP implementation of LIC1 by training (via the Backprop learning procedure) a randomised network to correctly compute some 10,000 samples of input-output pairs. However, I did glide over a few rough patches resident in this technology.

Notice to begin with that LIC1 is a function with no defined limits on its input values. This means that our training set, however large, cannot be the definitive look-up table for LIC1; it must be just a *sample* of possible input-output pairs. Hence, an MLP that successfully learns its training set may have learned any one of the innumerable functions that are consistent with the particular sample chosen, rather than learned precisely LIC1 (as we had glibly assumed).

It is true that the trained MLP can be tested with inputs specifically selected to rule out certain LIC1 variants, but because there are an infinite number of alternative possible functions, this strategy cannot be pursued exhaustively. We are left with the necessity to admit that the trained MLP computes the LIC1 function as far as we know, but we can never be certain.

Moving on to the process of training MLPs, although comfortingly well-founded, it is not all sweetness and light. Some MLPs refuse to improve significantly beyond an initial gain (i.e., they start to compute the training samples more accurately but then refuse to improve further). Such MLPs have become stuck on the top of a small hill, a "local maximum" such as we saw illustrated in the previous chapter as a sub-optimal peak. The MLP system is trapped on this peak because being a peak there are no further paths upwards. The only way off a peak is downwards, and a step down is a step to a worse system. If learning is set up, as the Backprop procedure is, to always deliver a step up, or possibly sideways, but not down, then there is no way off the top of a hill (but see GA systems below which tackle this problem directly).

Some MLP systems learn too well, i.e., they learn to compute exactly the training samples but fail dismally after training when given new inputs from the function that they have supposedly learned. These MLPs have been "over trained". This means that the feedback error correction of the Backprop procedure has in effect pushed the MLP to learn the training set pairs as the complete definition of the function rather than learn a general function for which the training set was a sample.

An MLP cannot discern user intentions, i.e., is the training set a complete function definition, or is it merely a sample? The human experimenter must be alert to this possible source of confusion. One safeguard against "over training" is to use two sets of input-output pairs — a training set and a test set. The former is used, as described earlier, to direct learning, and the latter is used to gauge the quality of the trained MLP.

Many of these difficulties must be solved by trial and error coupled with programmer experience. But suppose (as we did in earlier chapters) that we circumvent and overcome all these difficulties of MLP technology — what have we got? We have a trained MLP that implements the LIC1 function (to the best of our knowledge).

Further suppose that subsequently it becomes apparent that the function this network should be computing is not exactly LIC1, but a minor variant of it. Let's say that instead of the final "greater than" test (the ">" symbol in our earlier descriptions) we discover that what's really needed is "greater than or equal to" ("≥" as a symbol). This seems like a very small change, one that merely requires that we swap ">" for "≥" in the cognitive-level program for LIC1. But here with an MLP we run up against yet another negative repercussion of holism for the system-building engineer — there is no known way to make specific changes to the trained MLP so that it computes the desired variant.

It must be retrained from scratch, on the new training data. The holistic nature of the MLP computation (exposed in Chapter 6) strongly suggests that this complete retraining will always be necessary; it is not just a glitch due to our current ignorance[11].

Another limitation built into MLP technology is that the outputs are best treated as discrete, 1 or 0, pulse or no-pulse. So the functions that MLPs can be most easily trained to learn are functions with discrete outputs, and moreover a fairly limited number of different possible outputs. What are the practical consequences of this (not very well defined) limitation?

An MLP could not learn the ADDTWICE function, for example. The two input integers are not a problem, but how would you configure the MLP output units? One for each of the infinite number of different possible answers is clearly a non-starter. Maybe about, say, 11 output units to generate a sequence of 1s and 0s as (a binary) representation of the output integer. The largest number that could be represented is 2,047[12] which, although a good way short of infinity, isn't at all bad. But to learn this output interpretation requires that the MLP not only learns the correspondences between input pairs and correct output result, but it also has to learn the quite complicated "pattern" of this binary representation of a decimal number; it has to learn the "place system" that we use for number representation[13].

A much better way to approach the problem of training an MLP to learn this function is to transform it into a function that requires only a 1 (for "correct") or 0 (for "incorrect") output. So the input-output pairs become $m, n, p \rightarrow 1$ for samples in the function, and $m, n, p \rightarrow 0$ for samples not in the function. Consider the following table of input-output pairs.

$1,2,6 \rightarrow 1$	$21,3,48 \rightarrow 1$	$-3,3,0 \rightarrow 1$	$44,66,200 \rightarrow 1$	$17,23,80 \rightarrow 1$
$-10, -5, 30 \rightarrow 1$	$99,101,400 \rightarrow 1$	$167,13,360 \rightarrow 1$	$11,11,44 \rightarrow 1$	$0,4,8 \rightarrow 1$

These are all examples in the ADDTWICE function, if the first two input numbers are considered to be the two inputs, and the third is considered to be the corresponding output. For these "positive" samples from the ADDTWICE function the MLP will be trained to output 1 (as the table shows). But now we also have to include triples that are not in the ADDTWICE function. and train the MLP to output 0. So, $(21, 1, 48 \rightarrow 0)$ could be one pair from the "negative" training samples. In general we would train the MLP with roughly equal quantities of positive and negative samples[14].

The lesson of all this is that far from pointing the way to open-ended learning, MLP technology, which is by far the most successful of the general machine-learning technologies, only works well when restricted to functions whose outputs can be represented by something like a few 1s and 0s — i.e., a small selection of discrete alternative answers.

Let's just summarise the limitations of this most promising neural computing technology, limitations that must be borne in mind when anticipating its potential scope for use as a technology for simulating human learning:

1. It's limited to learning a single function.
2. To learn the function we must have access to a large number of input-output pairs — i.e., we must know the correct outputs for all

the training samples (which, of course, puts paid to prospects for open-ended learning).

3. It works best for functions whose output values can be easily represented as a small number of 1s and 0s.

4. We cannot understand the learned model at the cognitive level.

5. We cannot be sure exactly what function the learned model computes.

6. We cannot modify the trained network to compute a different function.

7. Optimal training regimes (e.g., mix of positive and negative samples as well as numbers of samples) are guesswork.

8. Over-training is always a danger to be guarded against.

All these limitations have been well known since 1981, at least. Yet towards the end of the 90s, Steven Pinker's popular book *How the Mind Works*[15] centres its challenge to AI systems on neural-network technology, or "connectionism". In 1997, Pinker still sees connectionism as the major technological challenge to traditional programming as the modelling technology for cognition. He invests a good chunk of his book laying out its fundamental inadequacies — the beating of a deceased horse is the gruesome picture that comes to mind.

Reality checks do not appear to operate when biological inspiration is in the air. Well into the new millennium, we can still read, "Since we do have several decades of experience in using self-organising paradigms, new insights from brain studies can quickly be adapted to neural-net experiments" (p. 270). On the other hand, maybe it's high time, for us at least, to let neural computing rest in peace, and move on to examine that other focus of naïve wonderment, Genetic Algorithms (GA) also known as evolutionary computation.

Recall that one way to easily picture the basis of most machine-learning technologies is as mountaineering or "hill climbing" in the argot of this computational specialism. The top of the hill is where the optimum system resides (e.g., the trained MLP that correctly computes a maximum of the test samples). The initial system (e.g., the randomised

MLP) is trying to get there from some sub-optimal starting point below the peaks.

Learning is thus a change in the system that moves it up the hill — a change to a more optimal (less sub-optimal) version of the basic system. For MLP technology (to continue with this example with which we have some familiarity), the learning is realised by link-weight changes, and a step upwards on the hill is a set of such changes that result in an improvement in computing the outputs for the set of test samples[16]. It is precisely the well-foundedness of the Backprop learning procedure that guarantees that every set of changes to the link-weights is a step upwards (or possibly sideways, or no-move, but never a step down).

This is the mountain range I've sketched previously, but now populated with MLP systems. It illustrates a piece of the space for different versions of the same MLP, i.e., same basic structure but different sets of link weights.

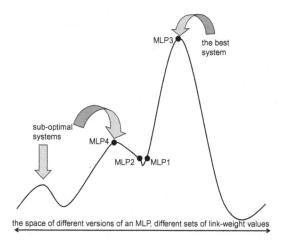

the space of different versions of an MLP, different sets of link-weight values

To reiterate: each set of link weights places our particular MLP somewhere on this curved line. Then MLP learning, which is changing link weights to give a better system, amounts to moving upwards on this curve. In the vertical direction we have quality of solution, i.e.

the better an MLP performs, however "better" is defined, the higher it is on a hillside. For the Backprop learning procedure "better" is defined as less difference between actual MLP outputs and correct MLP outputs.

In general, changes to the link weights of this MLP will move it on the curve — a change for the better will move it upwards and a change for the worse will move it downwards. Notice that it is possible for a link-weight change to have no effect on the number of test cases correctly computed, in which case the new version is no better and no worse than its precursor — so a horizontal move on the hillside (for simplicity, I've omitted the necessary plateaux). This sideways shuffling typically happens when the MLP is already on a peak, because it cannot move up and the Backprop learning procedure will never produce a worse MLP.

We can see that (within this sector of MLP space) MLP3 is the best MLP possible, because it is on top of the highest hill. It is the optimum solution, the weight-set that is MLP3 is the target for learning. Notice that MLP4 also sits on top of a hill, so it is quite a good solution but not as good as MLP3 (because it is not as high). MLP4 is a sub-optimal system — the Backprop learning procedure cannot adjust its link-weight values to get an improved performance. The smaller hill to its left, which doesn't happen to have an MLP sitting on it, represents the place where an even more sub-optimal MLP could end up after training.

When an MLP is initially set up with random weights, this amounts to a random placement of the MLP somewhere on this hillscape. MLP1 and MLP2 represent two such random initial MLPs, and being quite close together we might assume that their weight sets only differ slightly; they are very similar MLPs.

Because the Backprop learning procedure will never make weight changes that result in a worse MLP, every learning step will move the

MLP uphill (because this landscape has no flat plateaux, up, down or no-move are the only possibilities).

What we can easily see from this picture is that the two very similar starting MLPs — MLP1 and MLP2 — will train very differently. Assuming all goes smoothly, MLP1 will climb up the big hill to eventually become MLP3, an optimal solution. Similarly, MLP2 will climb its smaller hill to become MLP4. MLP4 is stuck at a sub-optimal solution, a local maximum, because the Backprop learning procedure (like virtually all machine-learning procedures) is set up to "learn", which is to say that it is designed to push MLPs up hills and never down them[17].

It is now easy to see (I hope) why hill-climbing as the basis for machine learning can get stuck at sub-optimal performance levels, i.e., the system starts learning, becomes a bit better but then stops improving far short of a good-enough system. What can be done?

Well, you and I can clearly see the problem with say MLP4. If we just back it down to the bottom of the valley where it started (as MLP2), and then move it across to the bottom of the big peak to the right, it can then be trained to ascend the biggest hill and so become MLP3. Simple. Even more simply, we abandon the local optima and adopt the (global) optimum, in this case MLP3.

What these simple analyses overlook is that in reality we (and hence our MLP systems) do not know very much about the hillscape within which the MLPs are moved by the learning procedure. In isolation, the lower peaks are indistinguishable from the highest one, so we never know whether or not a better system exists than the optimum that is first learned.

GAs present us with a way around this impasse of optimal systems that may be the best system available, but may not (they also present us with a few other innovations). What's the secret? Work with a population of systems, we'll call them model systems, or individual systems, to avoid confusion with the GA system as a whole. So-called "evolutionary" changes are introduced to jump individual systems onto new hillsides from sub-optimal peaks, and some poor individual systems are maintained in the expectation that they may carry forward innovative features that will be properly exploited later on, in future generations.

Maintaining a population of alternative model systems amounts to little more than distributing your eggs among several baskets (which is always a good idea if you have extra baskets). As an innovative insight it probably ranks below resisting the temptation to throw rocks at your neighbours if you live in a house with particularly large windows. There must be more to GAs. I confess to some puzzlement about the real value of this "more", but I'll give it all the positive spin I can muster in an effort to be fair.

The learning procedures in GA technology, the "evolution" of model systems, contain no guarantees, so changes introduced may result in worse systems as well as better ones. The absence of well-founded guarantees, such as MLP-learning offers, has the drawback of well, no guarantees when a GA system is set up to run — the outcome might be nothing useful, or it might be something surprisingly useful. But this does have the advantage of presenting us with a way to get off local hilltops. The individual systems in a GA framework are periodically subjected to major changes (i.e., they are "evolved"; details follow). This "evolution" step effectively flings the model system onto a whole new place in the landscape. This in itself would be nothing more innovative (and unproductive) than a new random restart unless there are good reasons to believe that the induced leap has a good chance of landing the "evolved" system on the side of a higher hill. The big question with GAs is: does it? Here's the GA-system picture (see next page).

I've used the same landscape, and the same four systems, although renamed "SYS1", etc. (they could easily be MLPs, but let's generalise

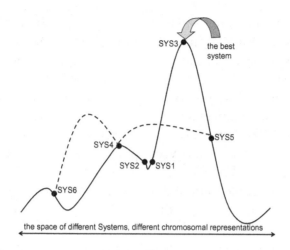

to model systems based on any learning technology). Each numbered "SYS" is a component of the population of whatever model system type the GA framework is being applied to. I've also added two new model systems and, as the dotted arcs indicate, both SYS5 and SYS6 directly derive from SYS4; they have been "evolved" from SYS4, to use GA parlance.

Neither SYS5 nor SYS6 are close to their precursor, SYS4, so quite radical changes have been made to SYS4 to generate SYS5 and SYS6. For example, if SYS4 was an MLP then the changes would not be small link-weight alterations (such as the Backprop procedure does), but big changes to link-weights, perhaps together with structure changes (deleting links, or even adding or deleting hidden-layer units[18]).

As mentioned earlier, radical changes to the make-up of a system will tend to fling the "evolved" model onto a new place in the landscape, which of course solves the problem of being stuck on top of a small hill (indeed on any hill). Two sets of "evolutionary" changes to SYS4 are illustrated, and both solve the local-maximum (i.e., low hilltop) problem. Or do they?

The new SYS5 can now climb all the way up to become the best solution, SYS3. But SYS6 has landed on an even smaller hill, and so can only climb up to a solution that is worse than SYS4. Such bad "evolutionary" moves are, however, to be expected and matter little in a population (they can be "killed off" or "evolved" again) provided that we get enough potentially good systems like SYS5. Although I've illustrated only six systems in this population, much bigger populations are typically used in a GA set-up. How many? Probably as many as can conveniently be "evolved", trained if MLPs, and tested within the resources available to the experimenter — just one of the arbitrary decisions to be made when setting up a GA system (see below).

What all this boils down to is whether the "evolved" changes to the model systems are no better than random (a bad thing) or are, for some reason, likely to generate potentially optimal systems (obviously a good thing). In terms of our illustration, we must get more systems like SYS5 than purely random changes would deliver. Otherwise, we might as well speed up and simplify the "evolve" step into a random shuffle, and jettison the claim to "biological inspiration".

For reasons that baffle me, the support for better-than-random changes is based upon the idea that biological evolution appears to have succeeded in producing optimal systems. So any imitation of evolution's techniques, however feeble, is also likely to deliver similar success. Leaving aside all the squabbling about what exactly is the essential basis of evolution's technique, let's examine the GA analogues.

To begin with, a system is represented as a string of (often) 1s and 0s. Whatever discrete coding is chosen, it should be obvious that innumerable linear codings are always possible for the essentials of any system. We must choose just one.

So our Chapter 2 brain-scan evidence based network, for example, might be coded as numbers that represent the hidden unit thresholds and further numbers (decimal ones, I guess) that are the link weights. Obviously, this linear coding is biologically inspired by the genome; it might be called a chromosomal representation.

Here is one possible chromosomal representation of our earlier network:

1	$w1$	$w2$	...	$w14$	$w15$	$t6$	$t7$	$t8$	$w16$	$w17$	$w18$	$t9$

Obviously the GA using such a representation must know how it is coded. In this case, the first element labels the system being represented — so this is SYS1. Then the 15 link-weight values in order are followed by the threshold values associated with the three hidden units (our original neurons 6, 7 and 8), and so on.

Further biological inspiration is introduced by implementing gene-like mechanisms to change these chromosomal representations, to "evolve" them in GA-speak. So sexual reproduction is mirrored by a "cross-over" mechanism in which two chromosomal representations are split and the two segments from one parent system are joined to the two from the other. In addition, some sort of random "mutation", i.e., random change to elements of the chromosomal representation, is also usual.

The following illustration shows use of a cross-over mechanism to "breed" SYS3 and SYS4 from SYS1 and SYS2. The point at which the parent "genes" are split (shown with a vertical dotted line) is another choice to be made, one that is only constrained by the requirement that the resulting offspring systems are viable systems (in accordance with whatever modelling technology is being used). The lower half of the illustration shows two possible mutations: one a point mutation at $w2$, and the second a segment mutation of the three threshold values in our evidence-based model.

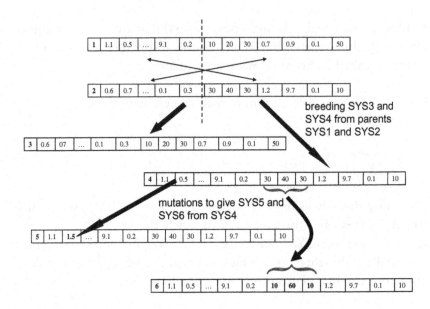

There are many further choices to be made, such as: what proportion of a population to "evolve"; how much and what ways to use mutation; which systems to discard; which to evolve; and which to leave unchanged from one generation to the next. A large number of more or less arbitrary decisions have to be made when setting up a GA system. This gives the experimenter great freedom, but little guidance on how to "set up a gene pool of possible software... [to] let the successful program *evolve* out of the soup"[2].

Let's assume that the mechanisms of cross-over breeding and mutation are central to the success of biological evolution, and that they do constitute better-than-random changes in biological systems. This latter assumption is required because biological evolution may have had the time to accommodate more or less random change, but the Singularity engineers definitely do not. Is this any reason to think that these biologically inspired mechanisms will similarly bestow direction to the quest for optimality in technical systems, mechanisms that can be no more than very faint echoes of the complexities of biology?

Why, for example, should an arbitrary coding of, say, an MLP as a discrete linear sequence mean that it then bears any fundamental relationship whatsoever to that of the genome in biological systems? "Biologically inspired" is the epithet used, but why on earth should we expect that such feeble "inspiration" is good grounds for anticipating a replication, or even just a faint echo, of the biological success story?

The tree of life from the earliest life forms at its base to the current diversity at its leaves is a similarly "biologically inspired" framework for organising the history of life on Earth. Does this mean that like its biological inspiration, i.e., the tree, we must anticipate that all current life forms will fall off (or die out as species) annually to be replaced after a period of "rest" by new versions? Or does it mean that we could expect to somehow feed the roots and so promote the proliferation of new species at the leaves? No. Why not? Because "biological inspiration" is no more than a basis for superficial guidance. It is not a licence to assume the full functionality of the biological source. Inspiration is not replication so any claim of carry-over behaviour from biological reality to "inspired" model needs to be justified.

In the absence of good reasons (and usually none are offered), mutation and cross-over are just complex ways of introducing random change. To have any plausible benefit within a GA system there must be a case made for why the swapping of system substructures, for example, when implemented by the particular cross-over details chosen, is likely to give better systems, rather than just randomly different systems — a useful number of SYS5-type systems rather than overwhelmingly SYS6-types.

To complete the picture of GA systems I need to tell you that after each evolution of the population of models, each model system is optimised according to the particular model type (e.g., MLP models will be trained with the Backprop procedure), tested against the criterion for success and so given a success, or "fitness" rating[19]. So with a population of MLPs, each MLP is subjected to a test set of input-output pairs, and its "fitness" is rated according to the percentage of correct test results. So "fitness" is GA-speak for how high up in the hillscape each individual system is found to be after modification and subsequent learning.

These relative fitness ratings are used to decide which model systems to keep unchanged, which to evolve (and perhaps how), and which to discard. Hence, we can achieve "survival of the fittest"!

We might, for example, evolve two new offspring systems from crossover using the two most-highly-rated systems, the two "fittest" individual systems. Then if we keep both the parents and their offspring, we have two extra systems to replace two discarded ones (perhaps the two lowest-rated ones, or two randomly selected low-rated ones — more arbitrary decisions required). By repeating this cycle it is hoped that some, or even just one, very successful system will emerge. Can it be more than a faint hope? Is it, rather, the sign of a suspiciously effortless breakthrough? Here's a picture of the GA framework:

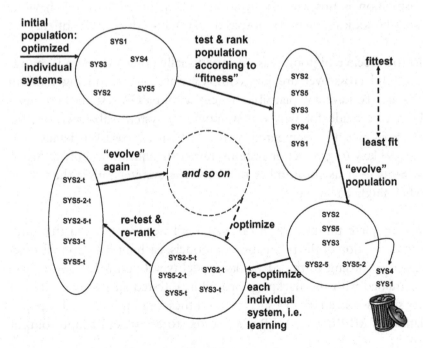

This is a completely automated process, so having set it all up, we stand back and, as Johnson says, we "let the successful program *evolve* out of the soup"[2]. The core steps — "evolve", learn, and rank for "fitness" — are repeated until a sufficiently "fit" version appears at the top of the ranked list, or until the population repeatedly fails to deliver any further improvement (in which case the experimenter must think again).

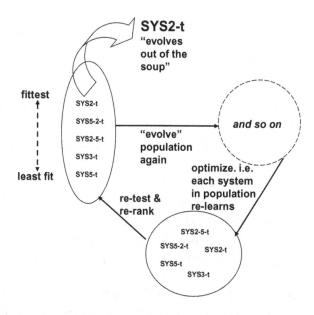

Sometimes the most "fit" system "that evolves out of the soup" is a surprisingly good one, but mostly it is not — the GA system churns around and around a population of mediocre model systems, and no really good one emerges. What do we do then? We fiddle with one or more of the many arbitrary decisions that we made in order to provide the necessary details for the crucial "evolve" step.

Working with GA systems can be particularly time-consuming and frustrating because of all these arbitrary decisions that are necessary (some are mentioned above). It should be obvious that the quality of each decision, whose implications take time to play out within the "evolving" essence of a GA, will not be immediately assessable. Even in the long run, how is the scientist to assess the quality of individual design decisions when they all interact in cycle after cycle?

Couple this with the absence of any guarantees of improvement of performance, and when good systems fail to emerge the experimenter is left wondering. Has it been evolving long enough? Are one or more of the many arbitrary decisions (e.g., exactly what proportion of systems to mutate) responsible for the overall unsatisfactory performance of the GA? Is the chromosomal coding causing the problem? Is an element of the "evolve" step (e.g., too much random mutation) responsible for lack of success? Is the function being optimised too complex for the systems being used? The possibilities are legion and any firm answers are hard to find because nothing is guaranteed and the various elements — coding strategy, evolution mechanisms, population-management decisions — interact in non-obvious ways.

Initially, I was puzzled to read that "a genetic algorithm does not accomplish its design achievements through designing individual sub-systems one at a time but effects an incremental 'all at once' approach, making many small distributed changes throughout the design that progressively improve the overall fit or 'power' of the solution" (p. 481). Although the individual systems are incrementally "designed" within, and constrained by, the overall context of the rest of the population, they are also "evolved" and tested for fitness individually. The "all at once" referred to is a new (for us) sort of holistic processing exhibited by a GA, but totally divorced from the holism we have seen in MLPs. Holism, as you now know, comes in a variety of forms.

This GA holism stems from the fact that decisions about which individual systems to "evolve" and how to "evolve" them are determined by properties of the complete population. There is nothing mysterious, and certainly nothing wonderful, in this feature of GAs. The fitness profile of the population is used to determine which to keep intact, which to discard and which to "evolve", so the details of the treatment of any one system is influenced by the relative success (and failure) of its co-inhabitants. For example, the two most fit (and therefore highest ranked) systems may be chosen as the parents of some number of offspring that then replace the same number of the least fit systems.

Although GAs are not holistic in the same sense as MLPs, the holism is almost equally unhelpful. In the case of GAs it means that when the total system is not working well — i.e., not moving smoothly from generation to generation to produce fitter and fitter model systems — it is close to impossible to pin down where exactly the blame lies, and therefore what needs changing and how.

Much earlier, I glibly explained that "the better an MLP is at correctly computing a test set of samples, the higher it is on a hillside". However, given that the MLP during training is learning to improve on more than one input-output pair (probably hundreds or even thousands), an "improvement" in computing the outputs for the complete set of training pairs is ambiguous. There are many different ways to define such an improvement. When training, for example, we might get the MLP to learn to compute correctly the first training sample, an input-output pair, and then the second one, etc. Alternatively, we might get the MLP to gradually improve its computation of all the training examples together.

With thousands of training samples, there are even more thousands of ways to define "an improvement" of the system, and hence a step up the hill. We must choose one. The Backprop learning procedure has, in effect, done this, and we know the choice is good enough because it comes with guarantees of attaining a hilltop. But when such formal guarantees are absent, as with GA "evolution" procedures, the user of the technology is left with yet another multitude of possibilities to choose between with little or no guidance (except experience from prior experimentation) about which choices are likely to be good ones.

The close analogy here is between thousands of training samples each of which can be individually pursued as a system improvement, totally

277

or gradually, (or the whole lot at once, or any subset), and a system that is learning to improve in terms of more than one behaviour. We might, for example, have a system that we want to be both more accurate and faster. After all, a very slow system that makes no mistakes may be useless. But do we try to get the most accurate system and then make it as fast as possible, or the other way around, or make small improvements each alternatively, or what?

Once the learning ceases to be improvement of just one type of behaviour then there are very many ways to define system improvement. Each choice amounts to climbing within a different terrain, and there are no easy ways to select a good landscape. "Try it and see", is the (im)practical strategy the GA user is reduced to when faced with this multi-behaviour, or multi-functional, optimisation.

The idea that an intelligent system will not be capable of many behaviours is ridiculous. The earlier illustrations of hillscapes are achieved with a single line because they are illustrating a single function (i.e., one behaviour, and thus a one-dimensional task) in the limited context of only one way to change the system (e.g., just link-weight changes for an MLP). Multi-function optimisation must operate in a multi-dimensional hillscape, and then "a step upwards" is a move that has a variety of different interpretations — adding yet another element of arbitrary complexity to a GA setup.

Let's ground this multi-function problem in the hill climbing that you or I might indulge in. Imagine you are standing on a real hillside rather than balanced precariously on the simple line that I've used in all the illustrations so far.

Upwards and downwards are concepts that you will have little trouble with. But you will be able to step upwards in a variety of ways, from a step up the steepest slope to a step up a gentler slope, with innumerable other options between the steepest and the least steep. And the choice you make can determine exactly what hilltop you will end up on.

278

In the following two-hill landscape, the hill climber at point "X" has many choices as to which step-up she will make. I've illustrated just two options. The leftwards step will result in a walk up the smaller hill and hence a final stop on peak A. But the step to the right will take you across to the larger hill and hence an eventually stop on peak B. The system on peak B is much better than the one on peak A, but which one is reached depends on a fortuitous choice at the outset.

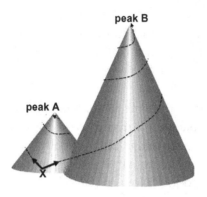

This wealth of possible upwards moves (learning steps for a computational system), with distinctly different consequences for how well the system will eventually learn, is manifest within a mere three-dimensional landscape. Multi-functional computational learning must operate in spaces of much higher dimensionality[20].

Okay, so we'll just use GAs to optimise the various single functions that constitute intelligence, and then put them all together. This is, as I'm sure you're only too well aware, yet another swing around what has by now become the well-trodden spiral into the modular-global impasse. Far from digging us out of the fundamental difficulties that the early AI enthusiasts encountered, these not-so-new technologies, as they stand, promise nothing — nothing particularly beneficial, that is.

"However, as is the case with all information-based technology, the complexity of genetic algorithms and other nature-inspired methods is increasing exponentially." (p. 482). Given the relatively fixed nature of our understanding of these technologies — i.e., recent decades have failed to reveal significant new potential — increasing complexity can only be a worry in a domain where unmanageable complexity is one of the major brakes on progress with AI-ish systems (indeed, with all IT systems). An exponential increase in complexity, if true, can never be a good thing, but in the absence of any significant expansions of potential, it spells nothing but absolute disaster.

In summary, it is not hard to appreciate the difficulties of the blind mountaineer climbing in an unknown landscape. This is the very difficult problem of machine learning.

But you and I learn all the time all sorts of things in all sorts of ways, it seems. It is true that your learning (mine too) is not perfect — you forget things, you learn things only partially, you even learn things wrongly. But all in all the human learning procedures (whatever they are) do a pretty good job. How come?

A good question. I wish I had a good answer. One part of the answer might be that evolution has had a good, long time to explore more or less random possibilities whereas machine-learning projects must hit on good results within time limits that are "immediate" when compared to thousands, perhaps millions, of years. Armed with intelligent analysis, the scientist ought to be able to extract the learning gems so laboriously discovered by evolution. But so far this has not happened. Maybe soon it will.

Endnotes

[1] Ray Kurzweil's *The Singularity is Near* (Duckworth, 2005), p. 444.
[2] Steven Johnson's *Emergence* (Penguin, 2001), p. 58. The author is transported into the realms of wild fiction by, what he calls, "bottom-up" technologies; in particular, his imagination leaps to the prospects for "open-ended learning" with a conviction that can only be sustained by ignorance of the technological realities.

[3] Jerry Fodor's *The Mind Doesn't Work that Way* (MIT Press, 2000), p. 47. This damning summary of the potential of neural computing as a technology for simulating the mind comes from an admittedly jaundiced philosopher. He has, however, debated extensively with the champions of neural computing, and (in my view) fully justifies his extreme comments elsewhere in the book.

[4] "I described how to apply the principles of evolution to creating intelligent designs through genetic algorithms." (p. 480) So said AI entrepreneur Ray Kurzweil.

[5] *Emergence* by Steven Johnson (Penguin, 2001) deals with "the connected lives of ants, brains, cities and software", and is interesting when the author sticks to the vagaries of ant colonies and cities, but he wildly overhypes the potential of the usual "novel" technologies.

[6] In *The Seductive Computer* (Springer, 2011) I try to deal with and defuse the hope, expectation and blind faith that we sometimes encounter in a variety of "novel" computational technologies, particularly those with "sexy" names. This definition of "emergent behaviour of software systems" is found in the Glossary, pages 309–310.

[7] The accurate and non-mathematical route to getting a grip on this style of "chaotic" computing is through an appreciation of chemistry and blacksmithing. In a red-hot bar of iron the atoms are all "jumping around" in a state of high activity (it's the heat input that causes this). If the red-hot bar is suddenly plunged into cold water (quenched), the sudden withdrawal of heat fixes many of the atoms in states that are not their lowest energy ones (local minima). They are trapped in relatively high-energy, but stable, states by the suddenness of the heat loss. Alternatively, we may slowly cool this red-hot bar down (annealing) to eventually reach the same temperature as the quenched one. The atoms in the annealed bar had ample opportunity to settle into their lowest energy states. Because of these differences in the states of their constituent atoms these two bars will have different properties, e.g., the quenched one will be brittle and the annealed one will not.

What's this got to do with neural networks? The weird name, Boltzmann machine, is because this technology uses Boltzmann's equations of thermodynamics. Boltzmann-machine networks are set up like a red-hot metal bar with a high "computational temperature" (just a value that fills the temperature slot in Boltzmann's equation) that causes activity values (the inter-neuron signals) to whiz chaotically between its constituent neurons. The networks are designed so that the lowest energy state they can settle into (a stable pattern of activity transfer) is the best result for whatever computation they were configured to solve. So problem solving amounts to reducing the "computational temperature" as slowly as possible but within a reasonable timeframe, called "simulated annealing".

[8] Within the experiments on the 30-city Travelling Salesman Problem (described and referenced in Endnote 9) repeated simulated-annealing runs of the network delivered "paths of less than 7 commonly, and less than 6 occasionally". This is the non-determinism, and it provides the experimenter with some guidance about the stability

and optimality of certain modelling choices, but little more than that. I find it most peculiar that commentators see such computational non-determinism as an echo of non-determinism in human intelligence and therefore an indicator that with this technology we're well on the way to AI, if not almost there already.

[9] At the considerable risk of being annoyingly repetitive (not to mention continual blowing of a trumpet that I happen to own), I can note that my *New Guide to AI* summarises and surveys the options in this class (called the "bath-like" networks) of neural computing. One interesting application of this technology, although not a Boltzmann machine, has been to produce "good" solutions to one example of that tiresome class of problems which we can prove that no computer will ever be able to fully "solve".

The game of chess is a well-known example. At most points in a chess game, no computer will ever be able to examine all possible next moves, and then all subsequent responses, and then all next possibilities, etc., because there are just too many possibilities in total (see Maxim 10 in glossary). It is simply this escalation of possibilities to be examined that give us a class of "exponential-time" (ET) problems. The fact that no computer can ever examine all possibilities and so find the best answer, does not mean that computers cannot be applied to these problems; it means that all such applications can only be relied upon to deliver sub-optimal answers. Computers can play good chess, great chess by everyday standards, but not perfect chess.

The so-called Travelling Salesman Problem (TSP) is another example of an ET problem. We imagine that the salesman has a list of cities to be visited and all he wants is the shortest tour (i.e., the order in which the cities are to be visited). Pretty simple given that he also knows all the inter-city distances, isn't it? Just write down all the alternative tours for the cities in question, add up the total distance that each tour involves, and choose the shortest. With six cities to be visited, our salesman merely has to deal with $6 \times 5 \times 4 \times 3 \times 2 \times 1$ alternative tours, which results in 720 possible tours to examine. This number of possibilities is quite manageable, but with only 30 cities to visit, the task becomes impossible for any computer. Multiply $30 \times 29 \times 28 \times ... 3 \times 2 \times 1$ and you will get approximately 10^{30} possible tours to be searched for the shortest. Assume one million (10^6) computers available, assume they each process one million tours per second (that's about 32×10^{12} tours per year), then the examination of all possibilities has been reduced to about 10^{11} years! This is hard to believe, but true. We humans exhibit a marked inability to properly appreciate the enormity of the very large numbers quickly generated by successive multiplications. A very readable account of these ET problems is given by D. Harel in *COMPUTERS LTD: What they really can't do* (Oxford University Press, 2000).

A Boltzmann-type machine was handcrafted such that its lowest energy states represented short tours for a 30-city TSP. After initial activation and simulated annealing, the tour represented by the final low-energy state was read off, and our salesman had a tour length of less than 6 (the shortest possible tour was believed to be of length 4.26, and the average length from 100,000 tours selected at random was

12.5 with none less than 9.5). The full details can be found in J. J. Hopfield & D. W. Tank's 1985 paper "'Neural' computation of decisions in optimization problems", published in *Biological Cybernetics*, vol. 52, pages 141–152, and a summary with some details is in *A New Guide to AI* pages 94–96.

[10] G. E. Hinton and J. A. Anderson stressed this "single-shot" limitation as long ago as 1981 in the book they edited, *Parallel Models of Associative Memory* (Elbaum; NJ, 1981), which also announced the major breakthrough of the Backpropagation learning algorithm, and hence launched MLPs as a well-founded and practical reality (although it should be noted that several other researchers make a similar claim).

[11] Very reasonably, you might note that the proposed change to LIC1 is a matter of simply adding another sub-function (i.e., the one that contains the equality test) to the trained LIC1 network. So why not just train this network a bit more with samples of the equality variant of LIC1? Can't we just "add" the extra sub-function instead of restarting from scratch? No, we can't. Further training with a sample of input-output pairs from the equality variant will utterly destroy the earlier training on the basic LIC1. This unhelpful phenomenon is known as "catastrophic forgetting".

[12] A sequence of eleven 1s interpreted as a binary number is $2^{10} + 2^9 + \ldots + 2^2 + 2^1 + 2^0$ which (to further exercise our recently-acquired rice-counting skills) equals $2^{11} - 1$ which is 2,047.

[13] The "place" system is the one we commonly use for numbers. In terms of 1s and 0s, it is a binary system in which, for example, 101 is to be interpreted as $1 \times 2^2 + 0 \times 2^1 + 1 \times 2^0$ which is $4 + 0 + 1$, i.e., 5. This is a complex output representation, and one that an MLP would be unlikely to learn in addition to the basic input-output relationship of the target function.

[14] This new approach is not quite the same as the original idea because it will no longer compute the output from two inputs; it will only compute whether the third input is, or is not, the correct output for the first two inputs according to the definition of the ADDTWICE function.

[15] Steven Pinker's *How the Mind Works* (Penguin, 1998) devotes one of his two largest chapters to refuting the connectionism claims to a virtual monopoly of working models of cognitive architectures. His view at the time was that neural computing components needed to be combined with traditionally programmed ones, and, in the absence of "hard" examples, that view is indisputable, but not very informative.

[16] Earlier we discussed MLP performance in terms of correct computation of training samples. Now that we are aware of the over-training danger, we'll estimate performance quality by means of correctness with respect to a test set of input-output pairs.

[17] The eager reader might observe that the big mistake is to start MLP2 on the small hill. Forget it and start with networks like MLP1 on the biggest hills. The big mistake here, however, is to forget that this illustrated landscape is *not* available to the experimenter. It is only an omniscient being that might have this information in view; the lowly technologists must work (pretty much) totally in the dark with respect to the landscape that the learning systems occupy.

[18] As I assumed that my simple landscape is the space of different link weights in the same MLP architecture, any such architectural changes (which are quite possible within a GA system) will amount to jumps within a multi-dimensional hillscape, and not just a repositioning on the single line illustrated.

[19] If the measure of "fitness" is to be degree of success on 100 test cases, as previously suggested, then any unevolved systems will carry exactly the same fitness rating from cycle to cycle, and this may be okay. But with a fixed set of test cases, the experimenter does run the risk of developing systems that are just very good at dealing with the specific 100 test cases being repeatedly used (an example of over-training). A better general strategy is to use new, random selections of both training and test cases on each cycle of the GA process. But note that the landscape our model systems are climbing within is partially determined by the specific test cases used (e.g., for some test sets a 100-system peak may exist and for some others, not). So, in addition to the climbing systems being blind in an unknown landscape, the topography changes constantly. This is not helpful, but if you're lost and can't see ahead anyway, perhaps it makes little real difference.

[20] You and I are limited to movement in three-dimensional space, but abstract mathematical systems can easily operate in "spaces" of whatever dimensionality is needed. The high-dimensionality of intelligence is manifest as all the "dimensions" along which it operates when exhibiting intelligent behaviours, e.g., speed (and faster is not always better), accuracy, appropriateness, courtesy... in general, suitability with regard to the many dimensions of the immediate context.

ೞ Chapter 12 ೮

Mind Recursion

"Recursion is the key capability identified in a new theory of linguistic competence... a 2002 paper [Endnote 15]...cite[s] the single attribution of 'recursion' as accounting for the unique language faculty of the human species."

Ray Kurzweil, 2005[1]

"It will always be true that our theories leave open innumerable questions about mechanisms."

Noam Chomsky, 1980[2]

"We suggest that FLN [our basic language faculty] — the computational mechanism of recursion — is recently evolved and unique to our species."

Marc Hauser, Noam Chomsky and W. Tecumseh Fitch, 2002[3]

There is a consensus among cognitive scientists that the human species jumped ahead of all the others over a period of time just a few tens of thousands of years ago — i.e., just "yesterday" in evolutionary timescales. This great leap forward was not in terms of our physical capabilities, but in terms of our mental abilities, our cognitive skills. The human species became intelligent. All other terrestrial life forms were left behind in the mire of a tooth-and-claw struggle for existence mitigated by some weak reflections of human cognitive skills.

In other words, full intelligence developed in the brains of the human species alone (although it's possible that some whales or dolphins might want to dispute this if only they could get through to us). It is this sophisticated ability to think and reason that sets us apart from all other occupants of the tree of life. It is, of course, exactly this uniquely human capability that science is struggling to understand, and that AI researchers would like to model and reproduce within a computer system.

The most obvious manifestation of our amazing cognitive skills is our ability to communicate with each other in terms of one (at least) of many thousands of human languages. Human language is open-ended. We, the practitioners, are endlessly creative when communicating with each other. Most utterances we generate and understand are unique, brand new creations within the unbounded wealth of possibilities that is, for example, the English language.

After listening to an interview with some whiz-kid football player, you might feel an urge to dispute the reality of this grand characterisation of the human language capability. However, the creative novelty of human language alluded to is mostly very low grade. It is not the rare flash of creative genius that gives the world new manifestations of narrative prose, enlightening poems, or breathtakingly original turns of phrase.

Everyday language creativity can be no more than the endless variety of word choices and minor variations in sentence structure that we both produce and accept, seemingly with very little effort. In stark contrast, all known animal communication systems appear severely limited, well bounded, and almost entirely lacking in similar creative novelty even at this basic level.

So this overt and easily explorable expression of intelligence, the human language capability, is justifiably viewed by some as a key to unravelling the complexities of intelligence. Indeed, this book began with an introduction to the Turing Test which is, in effect, a challenge founded on the assumption that human-language skills directly reflect intelligence. Any system that can talk the talk must be intelligent.

Here's an illustration that may go some towards justifying the detail that I feel compelled to run by you in order to expose the weaknesses in the usual explanation of the large question mark — i.e., what exactly did we humans acquire that suddenly thrust us way ahead of all other species in terms of language abilities?

How do we all accomplish this key skill — chatting? If we can crack this problem of explaining the human language capacity, we may be well on the way to understanding intelligence.

Well into this new millennium, a technical mechanism called "recursion" has gained fresh impetus as the preferred candidate that accounts for the uniquely human language capacity[4]. The acceptance of this viewpoint is now so well established that put-downs of animal communication systems can rest on showing that the species in question (from starlings[5] to chimpanzees[6]) do not process communications recursively, or that their linguistic productions are not recursive[7].

A further development of the grip of the "recursion hypothesis" is that explanations of the great leap forward, of the emergence of intelligence, can be boiled down to explanations of the "evolution" of a recursive processing capability within our cognitive architecture[8].

So what is "recursion" you might well ask? It is a way to define an infinite number of structures with a small number of rules. Probably not very enlightening, so let's look at an example.

The positive whole numbers, the integers 1, 2, 3, 4, etc., continue endlessly, do they not? There is no largest integer. If I give you a large number, say 100 trillion, then you can simply add 1, and give me back a larger number. In this case it would be 100 trillion and 1. In turn, I can add one more and give you 100 trillion and 2. Clearly there is no end to this exchange of larger and larger numbers, although there may be a few tricky moments when you need to know exactly how many zeroes there are in a billion or trillion. So, we have an endless, or infinite, number of whole numbers, and according to my description of recursion, by using it (whatever *it* is), we should be able to define this infinite sequence of whole numbers with a couple of rules.

Notice that every number in this endless sequence is equal to its preceding number with just one added on. Here's a rule that defines this characteristic of all whole numbers.

Rule 1: <a whole number> is <a whole number> + 1

> To be read as: every whole number is another whole number plus 1.

Maybe you can see that this definition captures the essential relationship between successive whole numbers as expressed above, but it does look worrying circular. So what's <a whole number>, you have every right to ask. This supposed definition doesn't seem to tell us, and it doesn't. But if we add one more rule, a rule that defines the first whole number, all becomes clear (I hope).

Rule 2: <a whole number> is 1

> To be read: 1 is a whole number.

Notice that Rule 2 defines the single exception to Rule 1, i.e., there is <a whole number> that is *not* another whole number with 1 added to it; it is the first whole number which is 1. Provided we always try to use Rule 2 first, and only use Rule 1 if Rule 2 fails to be applicable, this two-rule definition of the infinity of whole numbers works fine.

How has this small, additional rule sorted the circularity out? Let's look at an example: Is 3 <a whole number> according to our claimed definition?

Well it is not <a whole number> according to Rule 2, so let's try Rule 1. Ah, 3 is indeed 2 + 1. So according to Rule 1, 3 is <a whole number>, if 2 is <a whole number>. Well is 2 <a whole number>? Not according to Rule 2, but according to Rule 1, 2 which is 1 + 1, is <a whole number> if 1 is <a whole number>. Well, is it? Now Rule 2 finally tells us that 1 is <a whole number>. That means 2 is <a whole number>, and hence 3 is <a whole number> according to our two-rule definition.

Clearly this process can be repeated for our 100 trillion number, but it would take a long time. Nevertheless, it should be obvious that this process would work for any whole number. These two small rules define an endless sequence of positive whole numbers.

Together, the two rules tell us what <a whole number> is, and that there is no end to these whole numbers. With these two rules we can determine whether any proposed "object" is <a whole number> or not. Alternatively, these two rules enable us to generate or produce, any whole number within the endless possibilities, We simply start with Rule 2, which gives us 1 as a whole number, and then use Rule 1 repeatedly, which adds 1 to the previous whole number every time we invoke it, until the number we're seeking is generated.

So where's the "recursion" here? Rule 1 is a recursive rule. Why? Because it *defines a concept*, <a whole number>, *in terms of itself*. It is this circularity of definition that is called recursion, and that gives unbounded scope to our simple definition of the endless sequence of whole numbers. It is the simple, non-recursive, Rule 2 that saves the recursion from endless circularity. The simple statement that 1 is <a whole number> is our "stopping condition" for this particular recursive definition.

With "recursion" no longer just a strange word, we can return to the issue of the role of recursion in intelligence. Is it essential, probable, unlikely, or even detrimental? Are recursive mechanisms an element of the mind/brain, perhaps the crucial element that only humanity latched onto and so became intelligent? And should we be concerned about such an esoteric issue anyway?

As the introductory quotations illustrate, and we shall investigate further, attempts to account for the unique language abilities that you and I possess persist in raising this issue. Let me nail my colours to the mast — I think recursion here is a red herring and an interesting one, because it bears heavily on attempts to reproduce intelligence within a computer as we shall see in the following chapter. And this in turn carries implications for our quest to understand what makes us clever.

In the scientific study of human language it has long been widely accepted that defining a grammar is crucial[9]. The major function of a grammar is to specify the structures of the sentences that constitute the language under consideration.

So, "*The dog chased the cat*" has the structure of a sentence in English. It is grammatically correct, or just grammatical English. But "*The chased dog the cat*" is ungrammatical. It is not grammatical English. Why? Because it contravenes a rule of English grammar.

It is a collection of grammar rules that defines the first of the above example sentences as part of the English language, and the second as not. How many different sentences are there in English? Rather like the numbers given above, if I give you an English sentence, you can give me back a countless number of variations on this sentence — from minor word-order variations, through word substitutions, to any number of extensions with new adjectives, adverbs or whole new phrases.

Starting, for example, with *"The dog chased the cat"*, you can rephrase it as *"The cat was chased by the dog"* or *"The cat chased the dog"*. Both are grammatical English. You might choose to substitute any noun for *dog* or *cat*, even inanimate ones. So you might propose *"The dish chased the spoon"* — slightly odd, perhaps, but grammatical English nonetheless. You can add adjectives to get, say, *"The black dog chased the tabby cat"* and also add adverbs, *"The black dog vigorously chased the tabby cat"*. And so on[10].

No reader, I trust, is in any doubt that there are countless variations on this simple sentence, and there are, of course, countless other such simple sentences. This suggests that there may be countless different sentences in English. Given the finite length of time that you have to generate and understand sentences in your native language, the set is effectively infinite — you will never run out of possibilities for new variations.

The neatest way to characterise this infinite collection of sentences that is grammatical English is with a small set of rules, just like the two rules that define all the whole numbers. Recursive grammar rules appear to provide a means to satisfy this requirement. Does this therefore mean that your mind/brain can process recursively? Does it mean that human language, however managed in our heads, is a recursive language?

I shall answer "no" to both of these questions. There is no such necessity. Moreover, there is solid evidence against the supposed need for recursion (however loosely interpreted) as the fundamental basis of our language capacity. The need is illusory, and its realisation in language grammars is detrimental to boot. As you can see, the flag I'm flying here is the Jolly Roger.

The knock-on effect of the prevalence of recursive structures in these grammars is the consequent slippage towards the assumption that recursion must be a, if not *the*, central process in cognition. In order to see the problem of recursive grammar rules, we must dip into the intriguing complexities of human language and its relationship to recursive processes.

Your proficiency with the English language can be exercised in order to get to grips with the notion of recursion in grammar rules. How's this sentence?

The movie was applauded by the critics.

Just an everyday sentence that one might read, speak or hear without surprise. It can be expanded by embedding a "that" clause, such as *"that the script made"* within it:

The movie that the script made was applauded by the critics.

This is probably not a sentence that you have faced before, but still quite understandable, and firmly in the unsurprising category, nevertheless. Both you and I are unfazed by such totally novel sentences. It is precisely this ability that is the basis for the assertion of unboundedness — the potentially infinite possibilities of human language.

Let's beef this sentence up a little bit more by embedding another "that" clause, *"that the novel became"*:

The movie that the script that the novel became made was applauded by the critics.

This expansion produces another new sentence, one that teeters on the edge of understandability (and mostly falls on the side of incomprehensibility except for the most macho of native English readers who are prepared to bring a Herculean effort to the comprehension task). For no one does the meaning just "pop out" seemingly effortlessly, as it does with most English sentences we encounter in our day-to-day lives (all tough eggs can refer to the further embedded version in Endnote 11).

These three sentences are not very different — the third merely has "*that the novel became*" embedded in the middle of the second. Yet this small insertion, which mirrors the insertion that produced the second sentence, suddenly destroys our understanding (and similarly our ability to generate such sentences on the fly).

This incomprehensible sentence is not ungrammatical nonsense. Its structure is in accord with its predecessor; it merely contains one more embedded phrase than its readily comprehensible predecessor.

Is it grammatical English? This is a deceptively simple question. A common stance among the scientists who specialise in these things is to say it is "grammatical but unacceptable".

There is, however, no complete grammar for English because it, like every other human language, is not a fixed and universal language. It varies from place to place and evolves over time. One man's grammatical meat is another man's gobbledygook. We can, however, define a grammar for what we might loosely call core English, and stay within this core capability as we explore recursion.

So what is this grammar that admits such oddities? It is, you'll not be too surprised to learn, typically written to include recursive rules, and so capture the unboundedness of English sentence structures in just a few rules.

A grammar is no more than a set of rules that can be used to analyse or generate the "grammatical" sentences of a language in exactly the same way that our two earlier rules could be used to determine if a given "object" was <a whole number> (analysis), or used to produce any <whole number> (generation). In this sense the grammar rules define a language. They define the structures for all the grammatical sentences of the language; a grammar is often said to define the *syntax* of the language — the allowable structures regardless of meaning.

What the grammatical sentences *mean* is a problem of *semantics*, a knotty problem that we need not venture any further into. For us, for the moment, semantics is merely an issue of sentences we all readily understand, and sentences that none of us understand. In addition, this fundamental divide suddenly appears with no warning within a sequence of similarly-grammatical sentences — i.e., sentences all generated by the same grammar rule.

Readers who are endeavouring to build the big picture should perceive a link between our earlier struggles with the inadequacy of purely structural (i.e., syntactic) models, and our consequent quest for meaningful interpretations (i.e., semantics). Consider the value of a grammar for English with no meanings attached to any sentences. The fact that you only know exactly what word strings are sentences and which ones are not is (virtually) no help at all when trying to use this syntactic model of English for the purposes of communication which hinges on the transfer of meanings. This odd state of affairs is illustrated in the small story of the duped, and ultimately disillusioned, prisoner with which this book begins.

But back to the issue at hand: let's look at a very small grammar in order to clarify the possibilities and problems of recursion as a crucial element in the explanation of your amazing language skills.

We will define a simple sentence, called <sentence> with a few grammar rules. Sentences will be defined as noun phrases, called <noun phrase>, such as "*the movie*" followed by a verb phrase, called <verb phrase> such as "*was applauded by the critics*". Alternatively, a sentence may contain an embedded "that" clause, called <that clause>, such as "*that the script made*". A noun phrase is defined by **Rule 2** (below) as a determiner, called <det>, such as "*the*" or "*a*" followed by a noun, called <noun>, such as "*movie*" or "*script*". Verb phrases are defined by **Rule 3**, and so on, with **Rule 4** defining an endless possibility for embedding "that" clauses (as we shall see).

But first this small grammar[12]:

Rule 1 : <sentence> is <noun phrase><verb phrase> or
<noun phrase><that clause><verb phrase>
To be read as: a <sentence> is a <noun phrase> followed by a <verb phrase> or it is a <noun phrase> followed by a <that clause> followed by a <verb phrase>.

Rule 2 : <noun phrase> is <det><noun>
To be read as: a <noun phrase> is a <det> followed by a <noun>.

Rule 3: <verb phrase> is <verb> or <verb><object>
To be read as: a <verb phrase> is just a <verb>
or a <verb> followed by an <object>.

Rule 4: <that clause> is "that" <noun phrase> <that clause> <verb> or < >
To be read as: a <that clause> is the word "that" followed by a <noun phrase> followed by another <that clause> followed by a <verb>
or it is "nothing".

This fragment of English grammar is composed of four rules, each of which specifies acceptable grammatical structures for the entity named at the left end of the rule. So, to reiterate the basics before we dive into recursion: **Rule 2** defines a <noun phrase> as composed of a "determiner" (<det>, such as "*a*" or "*the*") followed by a "noun" (<noun>, such as "*movie*" or "*script*"). So "*the script*" is a grammatical <noun phrase> according to this grammar, but "*script the*" is not.

Rule 4 is the recursive one. Why? Simply because it includes what it is defining (the <that clause> which starts the rule) as part of its definition, i.e., the <that clause> also occurs on the right side of the "is". This dodge smacks of hopeless circularity. It is circular, but rather than "hopeless", it is one way to specify infinitely many structures on the basis of small finite means just as we saw with our "whole number" definition.

Rule 4 will suffice to generate (or analyse) sentences with any number of embedded "that" clauses, and, because they are generated (or analysable) using this grammar, they must be "grammatical". But as we saw with the three English sentences given above, after about two embedded "that" clauses the sentences are not comprehensible — we can't understand what they're saying, neither can we casually produce such sentences. **Rule 4**, though, sanctions any number of such embeddings as grammatical English. We will return to this oddity once we've seen just how our tiny grammar does get to grips with the infinite.

Rule 4 specifies two grammatical alternatives for a <that clause>. The first is the specific word "*that*" followed by a <noun phrase> (e.g., "*the script*") then followed by another <that clause>, and ending with a <verb> (e.g., "*made*"). So we know that a sentence fragment such as "*that the script*" <that clause> "*made*" is grammatical. We can reapply **Rule 4** to resolve the <that clause> in the middle of this fragment. If we use the first **Rule 4** alternative again, we might get "*that the script*" "*that the novel*" <that clause> "*became*" "*made*". Re-applying **Rule 4** once more, but using the second alternative of "nothing" for the <that clause>, we get "*that the script*" "*that the novel*" < > "*became*" "*made*". After removing the quotation marks and eliminating < >, we get "*that the script that the novel became made*", which is therefore grammatical. It is also (sort of) understandable by itself.

It is, as I hope you can see, the very existence of the somewhat odd option of < > (i.e., "nothing") as the second alternative for a grammatical <that clause> that saves the recursive grammatical analysis from endless circularity. The option of < > can halt the circular

recursive re-referencing of **Rule 4** at any point between no embedded
<that clause> and an infinite number of them; it is the "stopping
condition" for this recursive definition. Here's the second complete
sentence again:

The movie that the script made was applauded by the critics.

Let's process it with our (minimal) grammar. **Rule 1** tells us that this
sentence must begin with a <noun phrase>, and **Rule 2** tells us that
we have this with *"The"* as <det> and *"movie"* as <noun>. Back in
Rule 1, the next word in our sentence is *"that"* and this is not a <verb
phrase> but it fits the first option for <that clause>, **Rule 4**. This
matching proceeds smoothly with <noun phrase> as *"the script"*, the
<that clause> as "nothing" (the second option, < >), and *"made"* as
the <verb>. This leaves *"was applauded by the critics"* to be the final
<verb phrase> in **Rule 1**.

A two-dimensional representation of this analysis might help (a little):

```
rule 1: <sentence> is <noun phrase><verb phrase> or <noun phrase><that clause>
<verb phrase>
rule 2: <noun phrase> is <det><noun>
rule 3: <verb phrase> is <verb> or <verb><object>
rule 4: <that clause> is "that"  <noun phrase><that clause><verb> or <>
enter rule 1            The movie that the script made was applauded by the critics .
Is The movie a <noun phrase>?
   enter rule 2
   The movie is a <noun phrase>, exit rule 2
continue with rule 1
that the script made not a <noun phrase>, is it a <that clause>?
   enter rule 4
   that matches "that", so is the script a <noun phrase>?
      enter rule 2
      the script is a <noun phrase>, exit rule 2
   continue with rule 4
   Is made a <that clause>?
      re-enter rule 4
      made is not "that" but 'nothing' is ok as <that clause>, exit rule 4
      continue with rule 4, first entry
   Is made a <verb>? yes, that the script made is a <that clause>, exit rule 4
continue with rule 1
Is was applauded by the critics a <verb phrase>?
   enter rule 3
   was is a <verb>, and applauded by the critics is an <object>, exit rule 3
continue with rule 1
no more words and second alternative satisfied, so exit rule 1
```

This sort of grammatical checking is far from simple and easy, and this is not a book on the finer points of English grammar and parsing algorithms. So what do we need to take from the above exercise? We need to see where the recursive mechanism is used, and then we'll see what "machinery" is needed to make it work. Why? Because if your language skills really demand recursion-processing, then we know that somewhere in your head the essential "machinery" ought to be discoverable.

As noted already, **Rule 4** is the recursive rule, and you will see that it is used twice when checking the grammaticality of our sentence. What is crucial, though, is that the second time the processor uses **Rule 4** *it is already in the middle of using* **Rule 4**; this is the crux of recursive processing. It is emphasised in the diagram with the two arrowed structures, each of which marks the beginning and end of a use of **Rule 4**. The second use of **Rule 4** is *embedded inside* the first use. The processor starts using **Rule 4** afresh whilst in the middle of using **Rule 4**. It is this nested usage that appears to stress our conscious cognition, but is, we are told, the outcome of an essential component of our language processing skills and also a crucial element of what makes us clever.

If we now consider our third sentence, you will see from the analysis diagram that the further embedded "that" clause causes the appearance of another nested use of Rule 4. This is the sentence:

> *The movie that the script that the novel became made was applauded by the critics.*

And we get the following analysis according to our grammar:

The movie that the script that the novel became made was applauded by the critics.
enter **rule 1**
Is *The movie* a <noun phrase>?
┌─►enter **rule 2**
└─►*The movie* is a <noun phrase>, exit **rule 2**
continue with **rule 1**
that the script ... is not a <noun phrase>, is it a <that clause>?
┌─►enter **rule 4**, first entry
│   *that* matches "that", so is *the script* a <noun phrase>?
│   enter **rule 2**
│   *the script* is a <noun phrase>, exit **rule 2**
│   continue with **rule 4**, first entry
│   Is *that the novel* ... a <that clause>?
│   ┌─►re-enter **rule 4**, second entry
│   │   *that* matches "that", so is *the novel* a <noun phrase>?
│   │   enter **rule 2**
│   │   *the novel* is a <noun phrase>, exit **rule 2**
│   │   continue with **rule 4**, second entry
│   │   Is *became made* ... a <that clause>?
│   │   ┌─►re-enter **rule 4** again, third entry
│   │   └─► *became* is not "that", but 'nothing' is ok as <that clause>. exit **rule 4**
│   │   continue with **rule 4**, second entry
│   └─► Is *became* a <verb>? yes, *that the novel became* is a <that clause>, exit **rule 4**
│   continue with **rule 4**, first entry
└─►Is *made* a <verb>? yes, *that the script* ... *made* is a <that clause>, exit **rule 4**
continue with **rule 1**
Is *was applauded by the critics* a <verb phrase>?
┌─►enter **rule 3**
└─►*was* is a <verb>, and *applauded by the critics* is an <object>, exit **rule 3**
continue with **rule 1**
no more words and second alternative satisfied, so exit **rule 1**

Again, forget the details (assuming that there was ever a likelihood that you'd committed them to memory), just look at the arrows. Each arrow indicates the start of the use of a rule and the end of the use of that rule. The recursive processing of **Rule 4** can be seen as nested arrows, i.e., recursive processing means starting on another reuse of a rule before finishing the earlier use. Hence the individual usage arrows are nested one within another.

Contrast these nested arrows with the arrows shown for **Rule 2** and **Rule 3**. In both cases the rule is used, and the usage is finished with no intervening reuse. It is the recursive re-entry of **Rule 4** that demands a special mechanism, and it is this mechanism that humanity may have been blessed with and so leaped ahead of the other primates. So let's abandon all the grammatical nitty-gritty and focus on this special mechanism that might explain so much.

As noted above, the last example sentence (not to mention its analysis) has for most people fallen over the edge of ready comprehensibility. But for our grammar processor it is no more complex than its predecessor. It processes two recursive rule re-entries (illustrated with nested arrow structures). Such a nesting of "that" clauses (called "centre embedding"), just two deep, generally marks the edge of human comprehensibility[13]. But for a recursive processor, it just marks the very beginning.

You and I find this recursive style of analysis exceedingly difficult to follow, but for a properly equipped computer it is no trouble at all and could continue smoothly for any number of centre embeddings (but rest assured we will not). Let's just see one further step into the "grammatical but unacceptable" as this is the sentence that was studied some years ago. It is:

The movie that the script that the novel that the producer whom she thanked discovered became made was applauded by the critics.

Its grammatical structure, emphasising the centre embeddings, is:

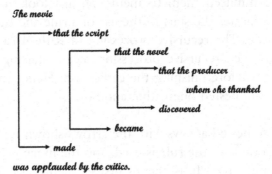

The problem for the human brain appears to be an inability to keep the sequence of noun phrases — "*the script*", "*the novel*" and "*the producer*" — in order to match each with its correct verb as they occur later in the sentence, in reverse order — "*discovered*", "*became*" and "*made*". But for a recursive mechanism this is all straightforward and easy.

300

In order to process recursively we must have a "stack" within our cognitive furniture; this is a storage mechanism that is found in all well-appointed cafeterias. Typically in such eateries, clean plates are placed on a sprung platform such that when a pile of plates is put in they sink down to present just the top plate ready at counter height for the first hungry customer. When this top plate is removed the pile rises slightly to present the next plate at counter height, and so on until there are no more plates left.

Notice that the plates are removed in the reverse of the order in which they were put in. The last one put in will be sitting on top as the first one to be removed, and the first one put in will be right at the bottom of the pile, and so be the last one removed. This is a classic "stack" mechanism, that's all it is, and it is just what's needed to process recursively.

In a stack the stored items are retrieved in exactly the reverse order from the order in which they were put in. The jargon talks of "pushing" items onto a stack, and "popping" them off again. So if "A", then "B", and then "C" are successively pushed onto a stack, the first item popped off will be "C", the next will be "B", and the last will be "A", then the stack will be empty. In the computer, as well as in the cognitive mind (I imagine), springs and stainless steel cabinets are dispensed with[14].

For a computer, a stack is no more than a last-in-first-out (hence, LIFO) storage mechanism. We can see its simple, but vital, contribution to recursive processing in the following illustration of the essentials of analysing the incomprehensible sentence according to our four grammar rules.

Each re-entry of **Rule 4** causes the current <noun phrase> to be pushed onto the stack. After processing the sentence up to "*whom*", the stack will contain the three <noun phrase>s in the order illustrated. After dealing with "*whom she thanked*" (a rule to do this is omitted), the analysis begins exiting from the multiple re-entries of

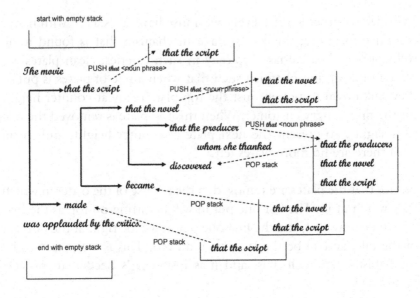

Rule 4. On each such exit, the top of the stack is popped to get the <noun phrase> that will be associated with the next <verb> in the sentence.

In detail, the first <verb> encountered is "*discovered*" and it is associated with the top item popped off the stack, i.e., "*that the producer*". Then the next <verb> "*became*" is associated with the next item popped off the stack, i.e., "*that the novel*". Finally, the last <verb> (in the recursive <that clause> rule), "*made*" is associated with the final item to be popped off the stack, i.e., "*that the script*".

Trying as this is for you and me, for the computer with a stack it is no trouble at all. It could similarly analyse sentences with hundreds of such centre-embedded "that" clauses. The stack mechanism ensures that any number of <noun phrase>s and their matching <verb>s will be correctly paired.

What are the essentials of memory structure and management for such recursive processing?

1. An ability to hold for the duration of sentence processing any number of phrases in the order in which they are encountered; and
2. An ability to access these phrases in reverse order.

This is certainly a neat and powerful computational technique, and it is proposed as crucial to the unique capacity of the human language faculty. Does that mean that it is the underlying mechanism, an important element of cognitive architecture? Some scientists appear to suggest this. "All approaches agree that a core property of FLN [i.e., the abstract linguistic computational system we all possess] is recursion... FLN takes a finite set of elements and yields a potentially infinite array of discrete expressions."[15]

Is the ability to process recursively the key to humanity's great leap forward? You and me, we somehow acquired an ability — essentially stack processing — that suddenly enabled our minds to manipulate boundless language constructions. Here's the picture that summarises the essentials of all the foregoing gruesomeness of grammars and sentence analysis.

Is the implication that agreement on recursion as a core property needed to define English entails the ability to process recursively in our cognitive architecture? Or is it just put forward as a plausible explanation with no implication that it is actually the mechanism used? For sure, a "property" is not the same as a "mechanism" but does this distinction hide any interesting implications, or is it just different words for much the same idea?

The chapter-opening quotations, taken at face value, suggest (if not assert) that a recursion processing mechanism is a component of our cognitive architecture. This assumption may turn out to be justified, and that would be interesting because in terms of conscious thinking, at least, recursive processing does not come naturally. It is a learned skill for computer programmers, and some never seem to be able to learn it despite becoming thoroughly proficient in the many other technical intricacies of software development. But any extrapolation from our conscious abilities to what might happen subconsciously cannot be relied upon.

In an important sense, recursion is, however, nothing special — any recursive rule and consequent process can be reworked as an iterative, or looping, rule and process[16]. Recursion is *never* a necessity, even for grappling with infinity.

Consider our "whole number" definition; it could be rewritten as a single "looping" rule:
<a whole number> is 1 with a number of 1s added to it

So if we add two 1s to the initial 1 we get the "whole number" 3; if we add a trillion 1s we get the "whole number" a-trillion-and-one; if we add zero 1s we get the "whole number" 1; and so on. I hope it is clear that this "iterative" rule can also handle the infinity of "whole numbers".

It is not recursive because <a whole number> is not defined in terms of itself such as was done in our earlier recursive Rule 1; it is a definition in terms of a simple looping process that adds 1s to an initial 1. Here's the picture:

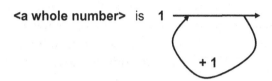

Similarly, the centre-embedded recursion in the <that clause> rule can be eliminated using two loops. The result, though, is visibly complex and ugly, but if you really want to inspect it, I've included it as an endnote[17].

Although both you and I may occasionally manipulate quite large whole numbers with ease, the same is not true of sentences with embedded "that" clauses. After just two such embeddings we are lost; we can neither understand nor casually generate such sentences. Given this limitation on our otherwise amazing language capabilities, no recourse to any sort of looping rules is necessary to explain this aspect of our facility with language.

Because we humans never deal in sentences in which the centre embedding goes beyond about three deep, a non-iterative reformulation of **Rule 4** could be:

Rule 4′: <that clause> is "that"<noun phrase1><verb1>

or

"that"<noun phrase1>"that"<noun phrase2><verb2><verb1>

or

"that"<noun phrase1>"that"<noun phrase2>"that"<noun phrase3><verb3><verb2><verb1>

To avoid any need for iteration, via either explicit loops or recursion, Rule 4′ simply lists the three alternative structures for an English "that" clause. In addition, to avoiding the complexities that recursion or loop iteration introduces, it has the distinct advantage of excluding any sentence with more than three centre embeddings. We would no longer need to plead "grammatical but unacceptable" for the incomprehensible sentences that we have seen in this chapter, or anywhere else for that matter. Adoption of **Rule 4′** in place of **Rule 4** means that one to three embeddings are grammatical, and four to infinity are ungrammatical. They have been excluded from the English language.

On the downside, our dealings with English sentences are not this clear-cut. Most English speakers will find three centre embeddings (such as our **Rule 4′** permits) incomprehensible, some will have the same trouble with just two embeddings, and the degree of difficulty we encounter with any such embeddings can depend on the specific words and phrases used. There are no firm limits that apply to all competent users of English regardless of any consideration of meanings. Hence, I suppose, the popularity of the neat and crisp, but unlimited, appeal of the recursive formulation as a specification of our general language "competence"[18]. However, notice that the "no firm limits that apply to all competent users of English" is a fundamental problem with any attempt to provide a grammar for many aspects of human language. It is not just a problem with embedded "that" clauses.

However, all English-language speakers find themselves limited to just a very few embedded "that" clauses. The fact that scientists have discovered no simple characterisation of this state of affairs does not mean that a recursive formulation, which is undeniably neat and crisp yet way over the top (with four to an infinite number of embeddings defined but unused), must be the explanation.

Indeed, it is a somewhat odd explanation, a powerful mechanism that we possess but barely use. Whatever the basis for our endless language creativity might be, we know, for sure, that endless embedding of "that" clauses has no part to play.

"Beauty and truth have a way of appearing to be akin."[19] Although the author of these words was focussed on the persuasiveness of a pretty woman, many scientists, especially mathematicians, are similarly smitten by good looks. They like to appeal to beauty, simplicity and symmetry as guides in their quest for accurate formulations of knowledge. However, nothing I know of requires that any of these niceties are indeed indicative of "truth". Indeed, some take the oddities and irregularities of natural systems to be evidence in favour of evolution shaping the system.

A well-known example is the "thumb" of the giant panda. This "thumb" is really a wrist bone that has evolved to fulfil the panda's need to efficiently strip bamboo leaves, its primary food, from the plant stems. Its real thumb, as with all bears, is non-opposable. Gould[20] discusses this particular oddity, as well as a number of others, and repeats with approval a quotation that this sort of non-elegant solution is "a contraption not a lovely contrivance". It would "win no prize at an engineer's derby" but such fixes are a typical outcome of evolutionary adaptation.

Nevertheless, the term "recursion", with its appealingly elegant formulation, used either somewhat casually or with technical precision, is typically invoked to explain our linguistic ability to generate infinite variety from finite means. But the recursive component of a rule such as **Rule 4** has no significant part to play in our endless capacity to both generate and understand new sentences. **Rule 4** defines the possibility of infinite embedding of "that" clauses, but you and I start stumbling on about three such embeddings. The infinity offered by such recursion you do not use. Therefore, the seemingly endless variety of our utterances must be due to something else.

This recursive grammar rule defines the wrong sort of infinity. How many sorts are there? The answer is disputed, but we need not dive deeply into this sea of contention.

Infinity is a tricky issue[21] especially when pursued to its variously weird and wonderful consequences. Luckily we do not have to try to make sense of the oddities to be found within this topic. We merely need to look at one or two different uses of the term, and appreciate the problems that they present for any recursive grammar rules designed to explain the endless creativity that you and I exhibit with our native language.

Infinity is a term that few of us have never heard, nor never used, but even fewer will be entirely comfortable with it either. Let's return to the "whole numbers" again.

For most of us infinity (denoted by a "lazy" eight — ∞) is the number at the end of all numbers. Or slightly more logically, infinity is the blurry state where the numbers are too big to count but just go on forever. An infinite number means lots and lots of whatever it is we're counting — an uncountable amount. It provides a convenient back stop for endless sequences. So the whole numbers are:

1, 2, 3, 4, 5, 6, 7, and so on to ∞

This is reasonably straightforward I hope, but notice that if we switch our attention to decimal numbers the above sequence becomes:

1.0, 2.0, 3.0, 4.0, 5.0, 6.0, 7.0, and so on to ∞

So what, nothing much has changed has it? It has. Consider this question: how many whole numbers are there between 1 and 4? Just two. And how many between 2 and 3? None, is the answer.

So let's move these questions to our sequence of decimal numbers: how many decimal numbers are there between 1.0 and 4.0? Quite a lot, aren't there? Let's run through a few of them in order: 1.0001, 1.001, 1.01, 1.1, 1.2, 1.2001, 2.11, 2.12, 3.09, 3.99999.

Choose any two adjacent numbers in this selection, say, 2.11 and 2.12. Are there any numbers between these two? There certainly are. Here are a few: 2.1101, 2.1111, 2.1112, 2.1119.

Choose an adjacent pair of these numbers, say, 2.1101 and 2.1111. Are there any intervening numbers? Sure, here are some: 2.110101, 2.110102, 2.1109. Each of these three numbers is bigger than 2.1101 and smaller than 2.1111. How many such intervening numbers are there? An infinite number.

To get to the point, the sequence of decimal numbers (just like the whole numbers) is infinite, it has no end. But between any pair of decimal numbers (unlike the whole numbers) there is an infinite number of intervening decimal numbers.

So, for decimal numbers we have an infinite sequence composed of an infinite number of pairs of decimal numbers that themselves each bracket an infinite sequence of intervening decimal numbers. We have an infinite number of infinite sub-sequences within the infinity of the decimal numbers.

That's confusing, but here's a pertinent simple observation: given an ordered sequence of decimal numbers and one new decimal number, there is just one place to put it in the sequence and maintain the ordering of the numbers.

For example, take our earlier ordered sequence — 2.1101, 2.1111, 2.1112, 2.1119 — and any new number, say, 2.11108. Where does it fit in? It is bigger than 2.1101, but smaller than 2.1111, so it fits into the ordering between these two.

Despite the confusing proliferation of infinities within the decimal numbers, every decimal number has its place in any ordered sequence of decimal numbers. Every decimal number is bigger, smaller or exactly the same as any other decimal number that you might choose to compare it with. How could things be different, you might well ask. And one answer is: when the objects of interest are sentences rather than numbers.

It is not uncommon for explanations of the "infinity" of human language to refer the reader to the more familiar world of numbers, just as I have done. Thus the 2002 seminal paper that has relaunched recursion mania, asserts that the infinity of the human language capacity is "a property that also characterizes the natural numbers" (p. 1571)[22].

But the infinities of human languages are not the simple infinity of the whole numbers (also called natural numbers) as stated above. They are much more like the complex infinities of the decimal numbers, and are in fact more complex. Sentences cannot, for example, be ordered according to length (i.e., number of words) because there are innumerable different sentences with the same number of words.

The sentences that constitute a human language offer no similar single and fundamental dimension of comparison. For numbers there is a single basis for comparison — size. Every number is bigger, or smaller or exactly the same size as any other number and, for numbers, that's about all there is to it. For sentences there is a rich variety of comparative properties, many dimensions along which sentences may be compared one to another. "Number of words" is probably one of the least interesting ones. Is "*A cat bit the dog*" bigger, smaller, or exactly the same size as "*The dog was bitten*"? The first sentence has more words but the second has more characters, and neither dimension of comparison touches upon their possible meanings.

Within the scope for unbounded creativity of language usage we find that recursive rules pursue one dimension — longer and longer sentences — and humans pursue an entirely separate dimension — endless substitutions and rearrangements.

Let's switch back to concrete examples of our language skills in order to see this fundamental distinction. Here are a few variations on a sentence that you've probably heard before. Most of these variations, though, are probably new to you, and some may be sentences that no one has ever contemplated before — novelties within the history of the universe. Yet none of my small selection poses a challenge to your ability to understand their meaning, and similarly you could continue adding to this set of variants effortlessly all day long. The main constraint would be boredom, not any sort of intellectual challenge.

This is the endless novelty and creativity of the human language capability. As these not-so-wonderful variants illustrate, quite a lot of different sentences constitute no more than a very small portion of your endless capability for novelty and creativity. However, this is not an undifferentiated mass. It varies along a number of different dimensions and we can pull out two significant ones.

The four arrows, each heading off to its own infinity, loosely illustrate the directions in which four different recursive grammar rules will create endless sentence variations. Thus the top-right arrow follows the grammatical productions of our old friend the <that clause> embedding rule. The top-left arrow charts the endless possibilities for adjectives to describe the "fox". The bottom-left arrow marks the path of endless qualification of the verb "jumped", and finally, the bottom-right arrow indicates the endless possibilities for adding a new sentence on to an old one by inserting "and".

Clustered loosely around the central "seed" sentence you will see all sorts of creative modifications. They result from all the word and phrase substitutions, additions or rearrangements that a huge vocabulary and a complex grammar make possible.

Along each of the four arrows something different is happening in detail, but all four can be viewed as treading the same path to infinity — one of continual growth in sentence length. In general, recursive grammar rules define a infinite sequence of longer and longer sentences. The language creativity they provide is founded on unfaltering expansion of sentence length.

Ever-increasing sentence length is a strategy that plays no part in human creativity with language. Rather we delete, insert or substitute words and phrases, and rearrange their orderings. Ours is a creativity founded on restructuring not inexorable growth in sentence length. These are two entirely different ways of approaching infinite variety of language use — two different dimensions of the infinite in language usage. The human dimension can usefully be called *structural diversity* distinct from the longer and longer sentences of recursion-generated diversity, *length diversity*.

In the following illustration, which is meant to be three-dimensional, we see length diversity in the plane of the page, and structural diversity heading off at right angles to this plane.

312

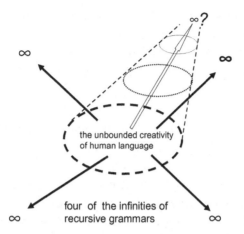

the unbounded creativity
of human language

four of the infinities of
recursive grammars

It is within this "tube" of structural diversity that human language creativity lies. As you can see recursive grammar rules shoot straight out of it as soon as a few (of the potentially infinite number) of recursive repetitions have been exploited.

To be more specific, the structural diversity within the central tube is the result of simple word order variations, minor grammatical rearrangement (e.g., active to passive), addition of words (e.g., adjectives and adverbs), and substitution of words (e.g., one noun for one of many other possible nouns). Such minor "fiddling" with sentences may seem pretty small beer compared to the elegant flights to infinity guaranteed by recursive grammar rules. Perhaps this is why such rules have generally been favoured by scientists devising a basis for the boundless creativity of human language.

The recursive grammar rules account for a diversity of language that humans virtually ignore. It is the wrong sort of diversity. It does not explain our language usage creativity. Then why is it such a popular choice? Perhaps because we humans (which includes many scientists) notoriously underestimate the explosion of combinatorial possibilities (as the Chinese sage exploited in an earlier chapter). That unbelievable truth (Maxim 10) exercises its malign effect once more. Combinations from the four categories of *structural diversity* mentioned above are quite enough to account for the unbounded diversity, or novelty (because many of the variations are quite new), of human language usage.

So if we jettison the wonders of recursion and recursive processing as the explanation of humanity's great leap forward, what are we left to cling to? The obvious candidate is the combination of word substitutions and additions as well as minor grammatical variations. It must be a very large, and ever-expanding, boundless store of words, a *lexicon* that can be combined in countless ways with a complex set of rules, none of which needs to be recursive or otherwise iterative (e.g., not even simple looping rules). Would that be sufficient to account for our endless creativity with language? One thing it wouldn't do is generate infinite variety. Hence the question mark next to the infinity symbol associated with structural diversity. We'll sort this out very soon.

In the case of human language, the grammar rules combine not only with each other but with a vast lexicon of, say, 50,000 words that is unbounded and continuously expanding. If the lexicon is viewed as fixed for a particular individual at a point in time, and the rules do not allow recursion or iteration, then the set of potential sentences is finite. Would this set be adequate to account for the *structural diversity* of language? Could it account for your apparently endless capacity to produce (or recognise) novel sentences?

As you might expect, not all scientists of language have been sold on the recursion argument[23], and similarly the infinity issue has been contested. Once we swap "boundless" or "uncountable" for truly "infinite", then many difficulties vanish. In particular, we don't have to worry about the "unattainable infinity" consequence of structural diversity. It is indisputable that neither you nor I will ever generate an infinite number of sentences (novel or not). Uncountable, maybe, and possibly boundless, but definitely not infinite.

In support of this simplification we find other commentators[24] also believe that the "infinitude of the set of all grammatical expressions is neither necessary nor sufficient to account for creativity of [human] language" (p. 126). Part of the argument presented by these scientists is built on the example of Haiku (and for us, Maxim 10 again), the highly structured form of Japanese poetry. Although accomplished Haiku poets are revered for their creativity their outputs are "selected" from a finite set of 10^{34} conforming utterances. To put this number in perspective physicists estimate that less than 10^{25} microseconds have elapsed since the big bang[25]. So, if we consider all Haiku poets that have ever existed, even collectively, they will not have examined every one of this finite set of possibilities. And to any one such poet, the possibilities are effectively endless, which is (I guess) where the creativity of composing a good Haiku comes in.

Do all these musings therefore mean that, to echo the exclusivity that typifies recursion claims, it is equally valid to assert that "a core property of FLN, our basic language faculty, is an ability to accommodate limited, explicit iteration, or possibly even just non-iterative alternatives"? I cannot see why not, and this claim (like so many other statements about how our minds work) is not saying very much at all. What it does perhaps do is switch the emphasis away from a fancy processing mechanism to a capacity to store and manage the combinations of a large number of simple grammar rules and a vast collection of words.

All this, even if totally true, does not mean that recursive elements would not play a part in useful models, simulations, of intelligence. "Biologically inspired", as we've seen, is a (dubious) "good thing" as far as the reverse engineering enterprise is concerned. So what are we to make of recursion that appears to be biologically shunned?

Way back in the 1960s, the psychologist George Miller[26] investigated the size of short-term memory (roughly: how many items can we keep in mind at the same time). He concluded that the number was 7 plus or minus 2, i.e., typically 5 to 9. In some lesser-known experiments he probed the human ability to comprehend English sentences with embedded phrases; it just so happens that one sentence he used was:

The movie that the script that the novel that the producer whom she thanked discovered became made was applauded by the critics.

Miller's experiments led him to conclude that everyone could handle one embedded clause, some could handle two, but everyone had trouble with three or more. So, comfortingly, the scientific conclusions about our inability to get beyond about three centre embeddings is an echo of our own, more casual, ones.

Is a stack mechanism, which is what recursive processing demands, limited to two or maybe three items really a stack? It could be nothing more special than the usual short-term memory facility within which we can manage to keep just a couple of things in order while we process a couple more items. Perhaps we then lose this ad hoc facility when the demand is greater than about two items.

It may well be the case that the term "recursion" is being used as a label for "some iterative capability". But do we need to be able to account for even this ill-defined facility? Suppose we shift the focus to language as used, "performance", rather than some chimerical "competence"? We are thereby spared the need to explain the almost totally unused potential for structural repetition inherent in recursive rules. Rules framed as a few explicit alternatives adequately define the language you use. They also eliminate the necessity for any complex iteration whatsoever.

Why is language "competence" (i.e., what we know about possible language structures rather than what structures we actually produce and can understand) favoured by linguistics specialists? Is it because it avoids the messiness introduced by getting to grips with both cultural and personal differences in our language capabilities? Perhaps so, but if mechanisms of our cognitive architecture is the primary concern, then a focus on what we might be capable of saying (if our cognitive architecture were different) rather than what this architecture actually generates and comprehends appears a little perverse. On the other hand, we do want to capture general cognitive mechanisms rather than the learned peculiarities of any given individual, or any group of individuals (although this latter knowledge could be useful).

As usual, it's a huge puzzle, and the current outcome as one leading researcher has put it in 2011, "I agree on recursion as one mechanism of iteration. What we lack is any sense of what the brain is doing when it comes to such computations. There must be some generative mechanism. Whether it is specific to language or shared across domains, we just don't know. In fact, most of the good stuff about language, we don't know much about it in terms of the brain!"[27] But, as argued above, the combination of a rich and fluid vocabulary with a few dozen (non-iterative) rules may be all the generative capacity we need to account for our seemingly limitless performance capability.

Using the term "recursion" (as a place holder for "some generative mechanism"), the above-quoted researcher and his co-authors[15] have provided an intriguing hypothesis to explain the uniqueness, within animal communication systems, of the human capacity for infinite variety. They suggest that the necessary capacity for recursion "evolved to solve other computational problems such as navigation, number quantification, or social relationships". Then "during evolution, the modular and highly domain-specific system of recursion may have become penetrable and domain-general. This opened the way for humans, perhaps uniquely, to apply the power of recursion to other problems" such as communication (p. 1758)[15].

This hypothesised explanation of the uniqueness of the human language capacity clearly has a bearing on the modularity of mind puzzle (the Chapter 4 topic) as well as the question of possible "saltations" that arose there. But we've undermined (at the very least) the assumed necessity for recursion within the human language capacity.

So what then, in general, accounts for humanity's great leap forward? If our language creativity is indeed a bellwether for what makes us clever, then it may be nothing more dramatic than memory size increase (to store the words, phrases and grammar patterns) coupled with memory-management enhancements (to accurately generate countless combinations of the various memory components)?

So here's another (and to my mind far more plausible) explanation of the large question mark in our original illustration of this problem.

Unillustrated is the memory capacity and management capability needed to systematically combine the grammar rules and word lexicon to both generate and analyse the boundless sentences of a human language. Do not forget that meaning, rather than grammatical correctness, is the fundamental property required of our utterances. So our explanation, however persuasive or not, does not address the primary purpose of language — i.e., the ability to carry and transfer meanings.

Finally, notice that with different phrasing, Miller's unintelligible sentence becomes:

She thanked the producer who discovered the novel that became the script that made the movie that was applauded by the critics.

It is a rephrasing that changes the emphasis of the sentence from the applauding of the movie to her thanking the producer. Does it change its meaning? Does the indigestible centre-embedded version have a coherent meaning? Not to me, but for a computer — no problem. (Well, no more of a problem than "meaning" in general for a computer.) Would the ability to process such deeply-nested sentences constitute an advance from natural-language processing to supernatural language processing? Is this our first inkling of the potential abilities of an ultra-intelligent machine?

Endnotes

[1] A quote from Ray Kurzweil's book, *The Singularity is Near* (Duckworth, 2005), p. 190 and derived from the article referenced in Endnote 15 (below).

[2] From Noam Chomsky's *Rules and Representations* (Blackwell, 1980), p. 112 in a section where he discusses what, if any, implications theories of grammar have for models of the human language capacity, and further he examines the evidence, both for and against, language being a recursive set — whether it is an infinite set of potential utterances that can be generated from recursive rules.

[3] Statement on p. 1573 of the paper referenced in Endnote 15.

[4] The 2002 paper referenced in Endnote 15 can be taken as the modern benchmark for the accepted status of recursion as the core of the human language facility. It has generated many commentary papers, which take issue with various elements of its claims, but they all accept a recursive basis of human language. A link between recursion and human language has been floating through the linguistic sciences ever since Chomsky proposed recursive components for grammars in the 1950s. The 2002 paper (which has Chomsky as a co-author) restates the relationship more forcefully and has become established as the basis for comment and criticism.

[5] Michael C. Corballis' article "Recursion, Language, and Starlings" in *Cognitive Science*, vol. 31 in 2007, pages 697–704.

[6] "Working memory: a cognitive limit to non-human primate recursive thinking prior to hominid evolution" by Dwight W. Read in *Evolutionary Psychology*, vol. 6, no.4, in 2008, pages 676–714.

[7] There is a good deal of confusion with respect to whether a language production, e.g., a sentence in English, can be said to be "recursive" (in general, this is incorrect). A large measure of this confusion is caused by loose usage of the term "recursion". We will enter into these intricacies only as far as is absolutely unavoidable.

[8] An example is *Recursion hypothesis considered as a research program for cognitive science*, Pauli Brattico in *Minds & Machines* in 2010, vol. 20, pages 213–241. Another example is *The recursive mind: the origins of human language, thought, and civilization* by Michael C. Corballis (Princeton University Press, 2011). He "argues that what distinguishes us in the animal kingdom is our capacity for recursion: the ability to embed our thoughts within other thoughts".

[9] Although, like most other interesting issues, belief in the fundamental basis of grammars is by no means universal amongst the leading lights in the science of this field. Yorick Wilks, for example, has long railed against the presumption that the human language facility is primarily syntax (i.e., grammar) based; he favours a more semantics-based approach to explaining the phenomenon, see for example Roger Schank and Yorick Wilks, "The goals of linguistic theory revisited" in the journal *Lingua*, vol. 34, published in 1974.

[10] Famously (in the world of linguistic sciences) Chomsky offered the sentence, "Colorless green ideas sleep furiously." as a grammatically correct, but meaningless, sentence. This is possible because human language does not admit the strict separation of grammatical correctness and meaning that we are implying. However, the complexities that would take our discussion closer to language-science realities do not have a bearing on the arguments presented, and so I omit them.

[11] *The movie that the script that the novel that the producer whom she thanked discovered became made was applauded by the critics.*

[12] Strictly speaking Rule1 does not need the first right-hand option, <noun phrase><verb phrase>, because it is covered by the second option when the <that clause> is "nothing" (i.e., the <> option in Rule 4). But this option is needed later when the non-iterative rule is substituted for the recursive one.

[13] This restriction (like so much else in the science called Natural Language Processing which, mercifully, is not our major concern) is not this simple. In 1982 a group of researchers, A. de Roeck, R. Johnson, M. King, M. Rosner, G. Sampson and N. Varile ("A myth about centre-embedding", pages 327–340 in *Lingua* 58), conducted an empirical study of the actual occurrence of such centre-embedded sentences which are assumed (by most theorists) to be exceedingly rare. They found

significant usage in certain German journalism as well as in at least one English writer, and so conclude that the limits of comprehension may be a learned skill rather than fixed language-processing machinery.

[14] The gleaming cafeteria plate-stack is necessarily of a fixed and pre-set size, i.e., the manufacturer's springs and chromium steel construction will take a fixed number of plates to fill it. It may contain less than this maximum but not more. A further advantage of the computational version is that it can be "dynamic", i.e., no pre-set maximum stack size is required because any stack in use can be stretched to whatever size is needed (within some ridiculously huge limit that any finite computer must impose).

[15] This statement from the journal *Science*, vol. 298, 22nd Nov. 2002, pages 1569–1579 in an article entitled "The Faculty of Language: what is it, who has it, and how did it evolve" by M. D. Hauser, N. Chomsky and W. T. Fitch is based upon the human-language property of infinite capacity from finite resources. In general, the authors are setting up a framework to explain how the recursive basis for this unique aspect of human communication might have evolved from other recursive capacities that originally evolved to solve other computational problems, such as navigation and number quantification. This paper provoked a variety of responses that all accept the central role of recursion in human language but question various of the original paper's other assertions — e.g., "The nature of the language faculty and its implications for evolution of language" by Ray Jackendoff & Steven Pinker in *Cognition* 97 in 2005, pages 211–225; and "Working Memory: a cognitive limit to non-human primate recursive thinking prior to hominid evolution" by Dwight W. Read in *Evolutionary Psychology* vol. 6, no.4, in 2008, pages 676–714.

[16] The absolute equivalence between recursion and looping (i.e., that they are always alternative possibilities) is believed to be true by many computer scientists, but I am not aware that it has been conclusively proven for all possible circumstances. However, for the relatively simple and straightforward uses of recursive rules in grammars for English, this inter-convertibility is assured.

[17] It also requires further iterative constructs (loop structures) as well as the arithmetic operation, addition, and temporary storage, called "verbs" below. Here's a non-recursive formulation of our earlier **Rule 4**:

Rule 4″: <that clause> is "that" <noun phrase> verbs=1 [a]while "that" verbs+1 <noun phrase> goto[a] for verbs downto 1 do <verb>

It is truly horrible to behold, and needs a lot of extra explaining, I would think, although we can now drop the oddity of the "nothing" option as there is no longer any need to halt the recursive looping. Our non-recursive **Rule 4″** looks slightly less awful when the non-linearities, the loops, are explicit.

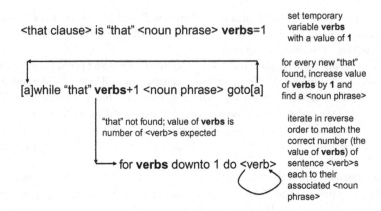

[18] Noam Chomsky championed the distinction between language "competence" (our "knowledge" of our language which the recursive rules are supposed to specify) and our "performance" (which the limited versions, we typically use, exemplify), see, for example, his classic *Language and Mind* (Harcourt, Brace & World, 1968).
[19] So wrote Ford Maddox Ford in *Parade's End* (BBC books, 2012), p. 853. He went on to say that a man "will go pretty far in the way of believing what a beautiful woman will tell him".
[20] In his collection of essays entitled *The Panda's Thumb* (Norton, 1980), S. J. Gould wrote, "The panda's true thumb is committed to another role, too specialised for a different function to become an opposable, manipulating digit. So the panda must use parts on hand [no pun intended, I think] and settle for an enlarged wrist bone and a somewhat clumsy, but quite workable, solution. The sesamoid thumb wins no prize in an engineer's derby." p. 24.
[21] Brian Clegg in his *A Brief History of Infinity* (Robinson, 2003) offers a history in the context of a popular exposition of this tricky concept. One of the examples of its trickiness is the infinite series:
$1, -1, 1, -1, 1, -1, 1, -1$, and so on to infinity. What is the sum of all these terms to infinity? With all the plus and minus 1s cancelling each other out, we cannot anticipate a large total, but what small number is it?
Well, let's bracket adjacent terms in the sum: $(1 - 1) + (1 - 1) + (1 - 1) +$ and so on to infinity. Each bracketed expression, $(1 - 1) = 0$, and an infinite number of zeroes added together will give a grand total of zero. So that's the answer, but wait, let's position the brackets differently to get: $1 + (-1 + 1) + (-1 + 1) + (-1 + 1) +$ and so on to infinity. Now we still have the infinite number of bracketed expressions each giving us a zero, but we now have a lone initial 1, so the sum must be 1.
[22] This claim is in the influential paper entitled "The Faculty of Language: what is it, who has it, and how did it evolve?" by M.D. Hauser, N. Chomsky & T. Fitch

published in *Science* in 2002, pages 1569–1579. The more complete quotation is, "All approaches agree that a core property of FLN is recursion... FLN takes a finite set of elements and yields a potentially infinite array of discrete expressions. This capacity of FLN yields discrete infinity (a property that also characterizes the natural numbers)." p. 1571

[23] In "Over the top — recursion as a functional option" which was published in 2010 in H. van der Hulst (Ed.) *Recursion and Human Language*. (New York, NY: De Gruyter Mouton, pages 233–244), P. Harder writes that, "Recursion may not be the right way to account for linguistic creativity." (p. 234)

[24] G. Pullum, & B.C. Scholz' article "Recursion and the infinitude claim", published in 2010 in H. van der Hulst (Ed.), *Recursion and human language* (New York, NY: De Gruyter Mouton, pages 113–137).

[25] A fact from David Harel's excellent book *Computers Ltd, what they really can't do* (Oxford University Press, 2000), p. 17.

[26] The source of this information is G.A. Miller, "Thinking Machines: Myths and Actualities", *The Public Interest*, vol. 2, pages 92–97, 99–108, 1966, and, more immediately, I have lifted the basis from an endnote in my *The Seductive Computer*.

[27] This is a personal communication (used with permission) from Marc Hauser, a joint author of the article in *Science* (Endnote 15), and sent on 7th February 2011.

☞ Chapter 13 ☜

Ultra-Intelligence

"A super artificial intelligence... won't have much effect on us because we won't be able to talk to each other."

Edward Fredkin, 1979[1]

"Bad news that is proven, lasting and robust... [is] problems that computers are simply not able to solve, regardless of our hardware, software, talents or patience."

David Harel, 2000[2]

Once science has a firm grasp of the mechanisms and principles underlying intelligence, improvements on the slow and error-prone human version will be forthcoming. Artificial Intelligence technology will soon leave our feeble ditherings behind. Science will create ultra-AI or super-AI. Will it? In answering this question, issues at the heart of what it might mean to understand intelligence are exposed.

Recall the "grammatical but non-acceptable" sentences in English that peppered the previous chapter. As was shown, judicious rearrangement can transform the unacceptable (in the sense of incomprehensible) sentence into a readily understandable one. Does this rearrangement change the meaning? Hard to say, given that prior to rearrangement the sentence did not have a coherent meaning for you (or me). Let's retrieve these two sentences:

The movie that the script that the novel that the producer whom she thanked discovered became made was applauded by the critics.

She thanked the producer who discovered the novel that became the script that made the movie that was applauded by the critics.

At the very least the focus or emphasis of the two sentences appears to be different — the first is focussed on the "movie" while the second switches to "she" who "thanked the producer". The point here is: what should a sentence that is incomprehensible to a human mean to a machine? What would the incomprehensible, embedded sentence above mean to an AI system souped up with stacks (or indeed any sufficiently powerful means) to process it recursively? Much the same thing as the comprehensible version, I suppose, with perhaps just the switch of emphasis pointed out above.

What will happen if the AI system starts to generate such sentences, which might require no more than a small exercise of its language-processing stacks according to the recursive grammar it contains? After all, why would an intelligent machine stop at two or three embedded <that clauses> when four, five, six or more represents nothing more than a little extra exercise of its recursion processing mechanism? Any human that is party to the conversation will be "lost" whereas any similarly powerful AI system will continue to converse smoothly, unperturbed by deeply embedded <that clause> sentences.

Our AI system will "cut-off" human intelligence through being endowed with a language-processing ability that does not contain the severe limitations that the human version exhibits. This is just one small, and perhaps in practice insignificant, ramification of doing away with the limitations of the human cognitive architecture. It is, however, but a first light on a huge and tricky issue.

The more that an AI system is built to circumvent the practical limitations on human intelligence imposed by our cranial wetware or the specifics of the cognitive architecture we have evolved, the more that the engineered artefact is likely to "inhabit" a different world (just like humans versus dolphins or whales, some believe).

It is true that an Artificial Intelligence could be explicitly limited to just three such embeddings, and so mimic human language usage. But such built-in limitations, even if they can be simply constructed without undesirable side effects, are contrary to the spirit of building an ultra-intelligent machine. (And if our machine is truly super-intelligent, it might well set about circumventing such arbitrary restrictions of its language capability.)[3]

The much-used distinction in language science between "competence" (all the sentences that can be generated from a grammar) and "performance" (the set of sentences that we humans actually generate or comprehend) would be transformed by our ultra-AI system.

In the previous chapter we saw that the recursive rule for defining embedded <that clauses> provided a short and snappy means to permit any number of such embedded clauses to be part of a grammatical sentence. Infinite possibilities from finite means are what a recursive grammar famously offers.

We also saw that neither you nor any of your friends can readily understand or produce sentences with more than about two or three such embedded <that clauses>, and two or three is a good deal short of infinity.

This somewhat esoteric line of argument was exposed previously in order to question the supposed necessity for recursive processing as an essential element of human intelligence. This view casts some doubt (at the very least) on the evolution of recursive processing as a unique attribute of the human mind, and hence an explanation of humanity's great leap forward.

We now pick up this thread again because it also bears directly upon the nature of an ultra-intelligent system; it is in this direction that we will now follow it. Time for some pictures, which are easily devised provided that once again we admit a degree of artistic licence with respect to illustrations of the infinite.

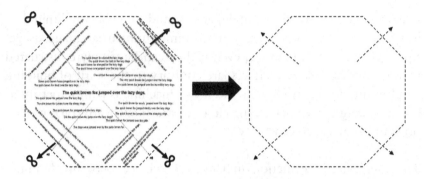

The left-side diagram of sentence variations taken from the previous chapter with its various indications of progressions to infinity is the collection of sentences that a typical English grammar might define as the scope of grammatical English. A boundary has been added to circumscribe this collection of possible sentences. As a nod towards the fact that this core collection is unbounded, I've dashed the boundary. Each of the arrows points towards a dimension of length infinity defined by a recursively-defined grammar rule (for simplicity, I assume just four).

We no longer need the detailed contents of this collection so I've simplified this (and the few other collections we'll need) to be merely boundary lines. This simplified version is illustrated on the right side.

Quite apart from our inability to get to grips with the infinities indicated, there will be a lot of simple sentences among the grammatical possibilities that every person will have had no cause to utter nor happened to have heard during a lifetime of chattering. Consequently, the collection of sentences that you or I (or anyone else) will get around to uttering or be required to understand during our lifetimes will be much smaller than even the dashed-line collection.

There are three such collections that we might consider: the collection of sentences that you happen to generate or understand in a lifetime; the much bigger, effectively endless, collection of all the sentences that you could have generated or understood should the

urge or need have arisen; and the collection that a grammar defines[4].
We'll forget the first of these — the happenstance collection of each
individual — and just focus on general possibilities for humanity and
machines.

When the possibility of Artificial Intelligence is entertained we obtain
a further alternative — the collection that can be generated or understood by the grammar and language-processing mechanisms built
into the AI system.

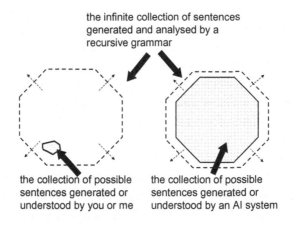

the infinite collection of sentences
generated and analysed by a
recursive grammar

the collection of possible
sentences generated or
understood by you or me

the collection of possible
sentences generated or
understood by an AI system

The simple point is that the collection of different sentences that you
and I may possibly process in a lifetime, although perhaps selected
from a countless pool, is a mere drop in the ocean of the infinite possibilities defined by the grammar. So the small irregular shape pictur-
ing this meagre collection — the scope for human performance — is,
in fact, illustrated as way too large. It should be perhaps just a microdot. Why this small?

Taking just the grammatical characteristic that we've invested some
considerable time and cognitive energy in — i.e., embedded <that
clauses> — we've seen that you and I generate and understand
about three embeddings, at most. So with respect to this embedding
aspect of our language, the small irregular shape circumscribing our

performance capabilities must represent sentences containing just none, one, two or three such embeddings.

But the grammar defines any number of such embeddings up to infinity. In the human collection we have sentences with three embeddings (at most), and in the grammatical collection we have an infinite number (again, at most). Three is barely a start on the long path to infinity. Given that the large dashed octagon with its arrows illustrates this infinity, then even a microdot is too big to illustrate our performance capabilities in relation to this grammatically-defined scope.

Mercifully, our super-AI system will be as unable to reach infinity as you or I (regardless of all the predicted exponential increases in basic technologies). In addition, within a human lifetime it will likely not generate or analyse many more sentences than you or I (unless, of course, large chunks of its time are spent chatting at supersonic speeds with other mechanical ultra-AI systems rather than communicating with humans).

But it has a much larger pool of possible English sentences to draw on, say, 300 <that clause> embeddings rather than your puny three[5]. Here's the picture of this mismatch in sentence choice.

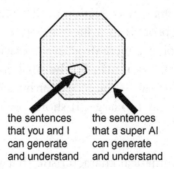

the sentences
that you and I
can generate
and understand

the sentences
that a super AI
can generate
and understand

What does all this mean for an ultra-AI project? Do we incorporate the neat, succinct and very powerful mechanism of recursion into its language and thinking capabilities, and then attempt to deal with the consequent communication problems? The resultant Singularity system may then be well named — as a system that can communicate with no one but itself.

Alternatively, do we try to copy more closely the human mechanisms? Do we build in the human sentence-construction limitations (limits due perhaps to the slow and unreliable nature of neurons). Do the scientists have to construct a language processing system that is more complex, messier and less powerful than a recursive one?

In a recursive grammar we see a clear choice point between an AI system with a human-level language capability, and a system capable of both uttering and understanding English sentences that humans are unable to manage — an ultra-AI with respect to language capabilities.

This ultra-AI will have no trouble, presumably, with the feeble human utterances but will have to cast around diligently in order to find a response that falls within the tiny "performance" set of the human. An ultra-AI system may have to be configured to talk down to humans, and to resist the temptation to exercise its full language generation capability. No such self-imposed constraints will be needed for one such machine to chat to another.

Just as we adults may adjust our utterances when talking to a toddler, so maybe a true ultra-AI system will of its own accord recognise the feeble communication skills of humans and adapt automatically when necessary. On the other hand, it might struggle to get communication with humans off the ground. We may become the whales and dolphins in a world run by ultra-AI systems!

What sort of "ultra" are we expecting or envisaging? It could be just faster processing because we've replaced slow neurons with very fast electronics, and done away with error-prone biology to give perfect information retrieval such as today's computers already deliver. So what might this ultra-AI be like?

With respect to its speedy information processing, it would, presumably, have to keep waiting for the humans to catch up with it. Perhaps not a big problem, the ultra-AI could switch to other matters while it waited, I imagine.

More substantively, what new ground might we expect this ultra-AI to break? Maybe faster processing implies no more than that it will get to its conclusions well before you or I do. If it remembers everything perfectly it will have more information upon which to base its conclusions, and so produce better ones than we typically do.

But note that the truly surprising and innovative ultra-AI conclusions will, by virtue of their surprising nature, strike us humans as odd, almost counter-intuitive by definition. Naturally, we humans when confronted with a surprising computer decision will wonder whether the unanticipated outcome is truly a creative innovation, or merely a blunder.

Precisely this issue quickly loomed large within the first expansive bubbles of the "expert system" euphoria of the 1980s. Was a diagnosis from a medical "expert system", for example, surprising because of new creative insights developed by the computer system, or was it surprising because it was plain wrong?

In order to address this issue a sub-field of "mechanical explanation" was developed alongside the main goals of automating human expertise. The idea being that a human user could query the "expert system" as to the rationale behind any of the "expert" pronouncements that were forthcoming. By so doing the human users could acquire confidence that a surprising AI-system outcome was not simply the result of a blunder.

As discussed at some length in the earlier chapter on "knowledge management", expert systems were anticipated in a euphoric wave of academic research and much launching of start-up companies. Initially, the early promise blossomed, but it eventually withered away in all the more ambitious manifestations to survive today only in a variety of severely limited realisations. Mechanical explanation followed suit. It worked quite well in small, logic-based applications but failed dismally in large-scale systems where much of the reasoning could not be based on clear true-or-false logical reasoning[6].

We can only assume that this need for "explanation" will re-arise with super-AI systems. And the new manifestation will be worse because the computer will be engaged in ultra-intelligent reasoning, presumably. In the context of large-scale, complex reasoning being performed by a system that has been built to circumvent the limitations of "human" methods, the challenge will be to describe the real sequence of reasoning in terms of the limited, human-level methods.

The alternative is to put absolute trust in the ultra-AI system, and accept its every decision without question. This strategy should send a shiver down every human spine. Let's leave it there.

More information means more searching for the relevant bits. Recalling the machine-learning adage "the more you know, the slower you go", the issue becomes one of whether the increase in searching speed is enough to offset the increase in the amount of information to be surveyed. In the absence of some quite miraculous breakthrough in data management (storage and subsequent access), which AI researchers have puzzled about for decades[7], will perfect memory and recall seriously undermine increased processing speed?

Forgetting, which is typically seen as one of the weaknesses of human intelligence, may be either failure to memorise, or inability to recall, or a bit of both. Is our propensity to forget a simple failing due to some fundamental indeterminisms of brain biology? Is it therefore something that modern, precise and reliable technology will banish from AI systems of the future? Or does it represent some optimal compromise between the occasional unfortunate lapse and a manageably small store of "knowledge"? We don't know, and even if this latter position is roughly correct, a much faster AI system could presumably work with a lot more stored "knowledge" than we do, forget less and still be faster.

IBM's Watson AI-system, which won the US TV game show *Jeopardy* early in 2011 and is dealt with in detail in a following chapter, initially achieved data retrieval times of about two hours (because of all the pre-processing needed to get "intelligent" answers to questions posed in English). By applying many more and faster processors to achieve massive parallelisation, the two hours was reduced to a few seconds. So hardware improvements (although not the whole story in the Watson triumphs) can, and will continue to, transform information management possibilities.

Once knowledge storage opens up to be far more extensive than the regular human capacity, issues of where and how to draw lines will come to the fore. As the major AI "knowledge storage" project, CYC (a Chapter 9 topic), found, there is no "end" to knowledge, and neither is there any definite form for its storage. There is always more that can be added to any piece of knowledge, and "pieces" do not turn up to just be stored. One fundamental choice is between storage as a set of facts (e.g., all the prime numbers less than 20), and storage of a rule that can generate these numbers. The storage of all knowledge, or even of all relevant knowledge, is a simple phrase that masks deep, underlying intricacies regardless of the existence of a virtually limitless, perfect recall, technology.

Is the knowledge that a super-AI system is based upon expected, or even required, to be totally consistent? I would guess so. It just doesn't seem right to have a super intelligence that holds contradictory views about some things. But your beliefs, just like mine, are almost certainly not totally consistent (as we discussed in an earlier chapter), but they are not wildly inconsistent either — they probably cannot be, if our consequent behaviour is to merit the label "intelligent".

Just as some degree of forgetting may be an optimal compromise between loss of important knowledge and the build-up of an unmanageably large amount (much of which will never be important), so the balance struck between consistency and inconsistency may be a similar optimal compromise strategy — one that ties into the "amount of knowledge" issue. Compromise as a key principle underlying intelligence will be developed further in the final chapter.

Maintaining a totally self-consistent set of knowledge and beliefs requires extensive checking of current knowledge and beliefs whenever any learning occurs. The compromise position might be to check only the "most likely" or "most important" areas of inconsistency. But then try defining "most likely" and "most important" for the super-AI system. It will be a thankless task.

Returning to faster processing alone (uncomplicated by an explosion in memory size), much in this chapter must be pure speculation, but not everything. We do know for sure that an ultra-intelligence, however fast it becomes, will never exhibit a perfect mastery of the game of chess, to pick just one small example from the swarms of so-called Exponential-Time problems that abound in our world. "A perfect mastery" means to always win when the opponent is not playing optimally and perhaps always forcing a draw otherwise.

How do we know this? It's a consequence of our old friend (Maxim 10): the unbelievable truth that **no computer will ever count 1, 2,... exponentially many**.

There are few certainties in this world, and even fewer where AI intrudes, but our inability to ever compute optimal solutions for the Exponential-Time, or ET, problems looks like one of the truly sure bets[2]. Why? The clincher is captured in the name: ET problems. The core difficulty is that the time needed to compute these problems increases exponentially as the problems get bigger, just slightly bigger.

Once more you should recall the surprising exponential escalation of rice grains that the wily sage requested from the Chinese emperor. In that example (which was one of surprising growth rather than out-right impossibility) the exponential increase was in amount of rice, but more generally the fundamental resource is time, the amount of time required to complete the task. A further small calculation will reinforce the utter impossibility of ET problems.

Let's look a little closer at chess. Computers can play pretty good chess already. The best ones will always beat most humans. Only the very best human players still have the edge, and even then they have to keep their wits about them as Garry Kasparov learnt in 1997. It therefore seems entirely reasonable to expect that an ultra-intelligent machine would easily trounce everyone. It may do this, but it will never play chess perfectly.

In general, within a game of chess, the computer (just like its human opponent) will need to carefully consider its possible next moves, then the opponent's possible counter-moves, then its possible responses to each, and so on. To play perfect chess it would be necessary to consider all these moves, counter-moves, counter-counter-moves, etc., to the point where the move-counter-move sequences end as a win, a loss or a draw. A well-known algorithm[8] could then be used to indicate which of the possible next moves the computer should make to guarantee a win (or, in the worst case, a draw).

Calculations of average move and counter-move numbers suggest that within a game of chess there some 10^{80} different board positions that the perfect chess playing computer will need to examine. As we have seen, this is short-hand for 10 multiplied by itself 80 times, and the small superscript number is called the *exponent*. It looks like a big number, but computers are unbelievably fast, and will get faster. So what's the problem here?

Well there are about 10^{16} microseconds in a century, and let's grant the computer the ability to process, say, a million board positions (10^6)

per microsecond, and we'll amass a million computers (another 10^6) to work on the optimum next-move problem. So the full set of move sequences will be searched in 10^{52} centuries![9] That's too long, so let's really beef up our resources. We'll assume we have a billion computers and they each can process a billion board positions per microsecond (we can't do much about microseconds in a century, I'm afraid). Now, what have we got? We've reduced the computing time to just 10^{40} centuries, and that's far longer than our universe has been in existence!

However we look at these ET problems — microseconds elapsed since the big bang, or number of atoms in the known universe, for example — it is quite clear that an ultra-intelligent machine is going to make only a little more headway than today's machines.

The crux of the difficulty is *exponential growth* in the size of the problem. In chess where we might have, say, an average of 25 possible moves from a given board position, each of which will allow our opponent to respond with one of 25 moves, and so on, the total number of board positions to be examined is $25 \times 25 \times 25 \times 25 \times \ldots$ for however long it takes to reach game end points[10]. So a short sequence of five moves and five counter-moves to a game outcome would involve 25^{10} board positions. Another sequence just one move longer would involve examination $25^{10} \times 25$ which is 25^{11} board configurations. Every additional step in the length of the sequence of moves adds 1 to the exponent. Hence we have an exponential increase in computational demands, and an exponential increase in resources (time, space, energy, etc.) can never be sustained for very long.

It is this exponential growth feature that is central to the inability to ever compute optimal answers to a wide variety of problems, and hence the general name for this class of intractable problems as ET problems. It could equally be, and sometimes is, Exponential-Space problems. The damning feature, in both cases, is exponential growth, which should give some pause for thought in those who see this feature as the basis for belief in super-AI "soon"[11].

Will the super-AI have a body like us, and live in the world like we do? Or will it be a largish cabinet that lives in a basement humming contentedly? I've no idea, but I can see huge problems with either option. As a disembodied "brain" the intelligent system will be established from the outset as alien — with either a pre-programmed, purely intellectual appreciation of what it is to be a human in the world (and consequent problems of updating this appreciation in the absence of all the feedback that you and I benefit from as part of the minute-by-minute interaction of our body with our world); or the system makes do with hypotheses about how we humans are different from it. In either case, this "intelligent grey box" option will be deeply problematic. So let's build a human-like body for our super-AI.

The development of a super-AI in the context of a human-like body as opposed to a disembodied grey metal box does appear eminently sensible if effective communication with humans is desired. The "body" then constrains and guides the directions of intelligence development. It provides a framework that should make any resultant intelligent activities somewhat human-like.

What this embodiment does not do is *lead* to human intelligence. The human body is the precursor of human intelligence, *but not the causal driver* (note: chimps, which in all bodily aspects are very like humans, have not made the great leap forward to sophisticated communication and thinking). Human intelligence developed to accommodate the human body; it did not develop because of the human body.

Quite apart from the direction of causal flow between bodies and minds, there is the further difficulty of how much, and what aspects, of the human body will be needed to sufficiently constrain and appropriately mould the work on intelligent processes. Clearly we cannot expect complete human embodiment (whatever that means, a *Blade Runner* "replicant" perhaps[12]), not at the outset anyway. So choices have to be made about what aspects of embodiment will be used, and what aspects (the vast majority) will be ignored. Is it a head, two arms and perhaps two legs that are most important, or perhaps a concentration on the head with two eyes, a nose and a mouth? The evidence from most robotics research supports these choices. But on what basis? Maybe teeth, a tongue, and a digestive system[13] is equally, or more, important as guidance towards an understanding of intelligence?

As usual, no one knows. But once embodiment is raised as important to the development of human intelligence, which is almost indisputable, it appears to licence whatever aspects of embodiment the experimenter chooses. Further than this, it is taken to imply that by working on, say, development of a grasping hand progress is necessarily being made towards the central processes of human cognition.

I see no simple reason for the robotics choices we are typically presented with (other than as a further trigger for our genetic imperative: **if it's like me, it's intelligent**). The cynic might observe that highly visible human behaviours, like shaking and nodding a head, are far better triggers of the genetic imperative than out of sight aspects of embodiment like sweet tastes and bad smells. But whether they are better staging posts on the long road to AI is anybody's guess.

Robotics work is very demanding, partly because the experimenters must grapple with the real world of imperfect objects and robot mechanics. It is also probably needed in order to contribute to the full package of an artificial human-like intelligence, one day. But its undeniable persuasiveness far exceeds its real merits as a step towards AI unless the embodiment choices are justified and the path from them to sophisticated cognition is, at least, outlined. Real science involves exploring general hypotheses. So what are these hypotheses when a robot is wheeled out to perform? All too often we are not told. We do not need to be. The genetic imperative ensures the necessary belief[14].

Robotics as an approach to embodied AI, from basic intelligence to super-AI, faces another manifestation of the fundamental problem of picking out the salient features (in this case, of a human body).

With only the human version of high-level intelligence to work with, we're a bit stuck. Much earlier, whilst endeavouring to reverse-engineer the underlying information-processing principles from brain-scan evidence (Chapter 2), we stumbled into the vortex of feature salience. The puzzling circularity of **understanding based a single example**. It is a non-resolvable circularity when a single example is all we have.

It's all very well to declare that the reverse engineer is aiming for "a duplication of the causal powers of the massive neuron cluster that constitutes the brain, at least those causal powers salient and relevant to thinking" (p. 463)[15]. But how can we pin down "salience" here? Trial and error? As our toaster examples showed, the failures of our models may indicate that we've neglected something important, but what exactly? And whether an identified failure point is the real root cause of the failure is likely to be an unhelpfully open issue. If we had just one other "model" of intelligence for comparison we might hope to make some headway with regard to this salience issue, but as it is, we're stuck on a treadmill.

Pinker, in his book *How the Mind Works*, offers us the "general principle... that any information processor must be given limited access to information because information has costs as well as benefits".[16] He lists the costs as space, time and energy (as examined above) although Kurzweil, for example, views these as non-problems because computer size, speed and energy demands are all moving exponentially in the right directions.

Pinker (p. 212) makes the further point (among many others) at length and with copious examples that intelligent vision is of necessity "solving" an ill-posed problem — i.e., the 2-D retinal image cannot contain enough information for unequivocal decisions about the shapes and substance of what is being seen. The human visual system circumvents this difficulty by various assumptions about the nature of matter and of the world within which it operates. One striking result, which we looked at in an earlier chapter, is the variety and persistence of visual illusions, examples of which are to be found in any book addressing the human visual system[17].

What do we expect of a super-AI with regard to its visual capabilities? Unless it has far more elaborate sensing equipment (such as laser range finding), it will also have to make do with incomplete information, and so be subject to similar visual "mistakes". And if somehow it were not, then there would be occasions when the super-AI and humans would "see" entirely different things when looking at the same scenes. Such a mismatch of perceptions again raises the issue of "mistake" versus superior judgement (in this case, with respect to what the super-AI claims to see).

It is unlikely that a super-AI will never be subject to visual illusion, and yet the reverse engineers are similarly unlikely to resist the addition of a few non-human sensors. Why omit the addition of, say, laser range finding when it's easy to add on, and can be decisively informative? Consequently, there will be occasions when the human and the super-AI system will have different illusions. Once again the scope for miscommunication between human intelligences and artificial ones looms large.

All purely technical approaches to the possibilities for an artificial ultra-intelligence throw up fundamental difficulties with the basic idea — i.e., that it might be possible even in principle, let alone in practice, to construct an intelligent system free of the many limitations we find in the human version. And beyond what I have called "technical" concerns, there is a vast field of softer issues where I fear to tread. I am thinking of consciousness, self-awareness, empathy, emotion, morality, etc., all elements (some might argue that they are crucial characteristics) of human intelligence. Although I've labelled these no-go areas the "softer" ones, we may well find that they generate the very hardest of barriers between the real and the artificial. But like so much else here, we really have no idea.

Singularity enthusiasts "solve" this problem by positing the ultimate system as some integration of the artificial and the evolved. Presumably the evolved component (i.e., you, me or better, much better, someone else) will supply the bodily-functions perspective and the artificial bit will add some amazing extras. That all sounds simple enough, apart from developing the AI component in the first place (no longer as a stand-alone intelligence but one that can somehow work closely and productively with an embodied intelligence), and then integrating it with the human with hopefully not too many "teething" problems.

As usual, the Singularity solution is on a par with that of the Sun-landing project that will land at night to avoid the obvious heating difficulties, and has already been taken to the top of a very tall building in broad daylight as the first step towards leaving the Earth en route for the Sun.

Glib expectation of phenomena such as the Singularity, which transcends human intelligence in all dimensions having shuffled off the undesirable bits, is premature, to put it mildly.

Endnotes

[1] Quoted by Pamela McCorduck in her book *Machines Who Think* (Freeman, 1979), p. 347, and attributed to AI researcher Edward Fredkin.

[2] From the Preamble in David Harel's *Computers Ltd; what they really can't do* (Oxford University Press, 2000), p. viii. A readily accessible popular-science presentation by a computer scientist on the known and proven limitations of all, and any imaginable, computer systems.

[3] The super-AI will of necessity embody powerful and open-ended learning capabilities so it could learn to overcome built-in initial limitations. If really super-intelligent, perhaps it would recognise the various "weaknesses" of the human-level intelligence foisted on it, and take steps to eliminate them.

[4] Just like everything else to do with intelligence, the situation with respect to language is nothing like this simple. Consider, for example, that many language utterances that you make or hear are not entirely grammatical. So, they're not in the collection defined by the grammar, yet they are in the collection of your personal

performance. This complicates the position but the essential argument can be made without recourse to this further complication (as well as a number of others).

[5] What will bound the recursive scope of an AI system? Stack size may be one limitation, although effectively unbounded stacks can be constructed that are only limited by the storage capacity of the total system (which may be unbelievably huge). Another limitation will be time — the AI-system must be able to generate sentences within some reasonable time frame, but, given the processing speeds anticipated, thousands of embedded <that clauses> must be a real possibility.

[6] Interestingly, a book by the doyen of UK AI, Professor Donald Michie, and a science journalist, Rory Johnston, proposed expansion of this "automatic explanation" facility as the saviour of complex computer systems. Their book, *The Creative Computer* (Penguin, 1984), was (as time has confirmed) just another example of the uncritical over-hyping of one aspect of AI — **by their** (lack of) **fruits you will know them**.

[7] AI researchers (and all IT-system developers) tend to see their systems slow down as more and more information is added and needs to be considered en route to better outcomes. Sorting (i.e., organising the ever-increasing information) and searching (i.e., finding the relevant bits in an ever-growing collection of information) are huge topics in computer science because of their importance in so many IT systems, and a super-intelligent system is not going to dodge this difficulty.

[8] It is called the "minimax algorithm" and can be found in any computer science textbook that deals with computer game playing.

[9] We start with 10^{80} board positions to examine. With a million, 10^6, examined every microsecond and 10^{16} microseconds in every century, we'll have $10^6 \times 10^{16}$ board positions examined every century. That's $10^{(6+16)}$, or 10^{22}. So we expect the 10^{80} board positions to be examined by our super-fast computer in $10^{(80-22)}$ centuries. That's 10^{58} centuries. But with a million computers working in parallel we can cut this down further to 10^{58} divided by 10^6, which is $10^{(58-6)}$, giving us just 10^{52} centuries to wait for the perfect move to be computed! With some time on his hands while waiting, the programmer might note that there have (according to Harel, Endnote 2, p. 17) been less than 10^{25} microseconds since the Big Bang.

[10] In David Harel's book (Endnote 2, above), which explains many quite surprising limitations on all conceivable computers, he estimates an average of 35 alternative moves for any chess-game board configuration, and 80–100 moves and responses in a typical game. So my estimates are on the low side, but that makes absolutely no difference in the context of such huge numbers.

[11] It is of course Kurzweil's Singularity book that makes so much "virtual hay" from his vision of exponential growth everywhere.

[12] In *Blade Runner*, the 1982 American science fiction film directed by Ridley Scott, at some point in the future, robotics technology has developed to the level where it is difficult to distinguish the robots from the humans on the basis of overt behaviour; these almost-human robots are called "replicants".

[13] Human intelligence is obviously well adapted to the human body — ingestion of beneficial substances, such as fats and sugars, is reinforced by their pleasant tastes, while bacteria tend to smell (a precursor of taste) bad so that we seldom get to the point of ingesting such harmful substances. Does this mean that the human body evolved to make positive use of tasty substances, and to reject unpleasant ones? No. It means that taste sensations, from sweet to repulsive, evolved (via survival of the fittest) to deal with potential ingestion of available substances from beneficial to damaging, respectively. Modern humans suffer from many of these evolved sensations because the context has changed. Throughout our evolutionary history, sugars, for example, were scarce and consequently their ingestion needed to be maximised. In the modern world, sugars are cheap and plentiful in many societies and we witness these human predilections killing us through, for example, diabetes and more generally, obesity. So to work from human-like bodies to develop human intelligence is like working from horse carts to develop horses — a classic case of putting the cart before the horse.

[14] There are, of course, exceptions but well-founded theories justifying the role of the chosen aspects of embodiment are never explained and probed in the course of media presentations which cannot get enough of robot cameos as entertaining "science". As a counter-example, Angelo Cangelosi at the University of Plymouth in the UK is working on an iCub robot (featured in Chapter 1). He is exploring hypotheses about how having a body and a sense of placement in space together with seeing something and not seeing it might be necessary for (and so explain) the human toddler's ability to build a meaningful linguistic framework for the objects it encounters in the world. See A. Cangelosi (2010), "Grounding language in action and perception: From cognitive agents to humanoid robots" in *Physics of Life Reviews*, vol. 7(2), pages 139–151 for a review of language and embodiment modelling, which includes the iCub example.

[15] Earlier in Kurzweil's book (p. 443), "we find that our ability to replicate the salient functionality of biological information processing can meet any desired level of precision".

[16] Steven Pinker's *How the Mind Works* (Penguin, 1997) p. 137.

[17] Richard Gregory's books *Eye and the Brain* (5th edition, Oxford University Press, 1997), and *The Intelligent Eye* (Weidenfeld & Nicolson: London, 1970) are two good sources of illusory pictures as well as discussion of the nature of this important component of human intelligence.

ᘓ Chapter 14 ᘒ

Semantic Mirages

"The word 'means' is constantly used so as to imply a direct simple relation between words and things, phrases and situations. If such relations could be admitted then there would of course be no problem as to the nature of 'Meaning'."
C.K. Ogden and I.A. Richards, 1938[1]

"The problem of meaning is of major importance in the study of the nature of intelligence."
Robert K Lindsey, 1963[2]

Mirages are real. I've seen them clearly in the Sahara desert. At least, they are as real as other perceptions when viewed from a distance, but as you get closer they dissolve away. At the time, we were searching for the next oil drum that marked our route across the desert. It was then disconcerting to see an array of oil drums on the horizon, but with binoculars we soon sorted out the visual fictions from the single reality — and we had the heading we needed. If only we could find the "binoculars" to similarly strip out the true semantics from the plethora of possibilities that shimmer and dance around the periphery of our cognitive models, or worse, fail to put in an appearance altogether.

The "meaning", even worse the "real meaning", of actions and processes observed from a distance often promises a similar reality, but once up close the promise dissolves, only to be reconstituted once more, at a suitable distance. So goes the AI-quest for meaning at the cognitive level, the *semantics* to support AI. We've glimpsed this problem already in the quest for a cognitive-level explanation of the brain module that we explored in detail at the outset of this book.

In the early days of mass AI (beginning in the 1960s), the market was awash with books that had "semantic" in their titles, such as *Semantic Information Processing*[3]. Everyone in the business knew meanings were crucial, and they hoped that they might be made manifest by the augmentation of their fancy structures and information-processing algorithms with "semantic tags", "semantic markers", or some such "content capture" strategy.

On another line of attack, knowledge representation was founded on "semantic networks". We've already explored a few semantic networks as part of our earlier efforts to get to grips with what exactly iCub "learns" when presented with a purple car and an octopus.

The Pandora's box of multiple meanings was first flung open when AI pioneers realised, and publicised their insight, that computers were not just adding machines, number crunching devices. Numbers, so the realisation went, were just a special case of symbols with meaning in number systems — integers, decimal numbers, binary numbers, etc. Computers were symbol manipulation devices, and the symbols could mean whatever we want them to mean — the meanings attached to computations was our choice. Why limit it to numbers?

Right at the beginning, I kicked off this polemic with a text-bite hijacked from the anti-AI philosopher, John Searle: "Syntax is not sufficient for semantics." Less pretentiously, this basic idea was summarised as Maxim 4, and variously applied as the undeniable truth that **structure does not provide meaning.**

I maintained that these five words encapsulated the fundamental reason why the strategy to reverse-engineer an understanding of intelligence from brain-scan evidence is fatally flawed. In the interests of initial simplicity, I did nothing to dispel the implication that there was a clear distinction between structure and meaning in our efforts to extract the makings of an understanding of intelligence from close examination of the workings of a brain module.

Subsequently, we have seen that this simple two-part interpretation is a fiction. Consequently, I had to tie my argument down to arrangement and function at the anatomical level as the (potentially) derivable structural view, and to cognitive-level behaviour as the required meaning. This latter view, the necessary semantics for engineering an AI system, was particularly difficult to pin down because we know so little about the structures and mechanisms of our cognitive architecture. We conjectured in general about localised, compositional structure as well as explained and explored holistic possibilities. In particular, we focussed on the cognitive-level meaning of the brain module under investigation, i.e., the details of function LIC1.

Although not as nebulous as "semantics", the structural view is neither neat nor simple either. In sum, there are multiple levels of both structure and meaning.

As we saw, the brain module has an obvious structure in terms of the component neurons and how they are connected together. Then the information-processing principles that we extracted (with a good deal of luck, inspiration, and a little cheating) might be termed the meaning of this structure. It was, however, not the meaning that we needed in order to understand the intelligence that the brain's neural network supports.

If we push on down to molecular and atomic levels, we can view the structure of a neuron in terms of interconnected and interrelated molecules whose meaning is best captured by the movement of positively and negatively charged atoms and molecular fragments. This low level of meaning is more usually presented as the process of a neuron receiving and processing incoming signals from other neurons.

At an even lower level, each of these molecules has an atomic structure. This can be explained and understood in terms of protons, electron configurations, inter-atomic forces, etc., and so it goes on, level to level (well probably not much further downwards).

The structure extracted and explored in an effort to relate it to, and so explain, meaning is level-dependent at both ends of this hoped-for bridge, and maybe the choice of levels is crucial to our chances of success.

For example, we would not expect to explain the functions of DNA if biology had focussed entirely on quantum mechanical interactions among elementary particles. "Inheritance would not have been understood. The level of explanation would have been wrong. Quantum mechanics lacks a notion of function, and its relation to biology is too complex to replace biological understanding. To understand biology, one must think in biological terms."[4] Just as neuro-anatomical processing principles (such as brain scanning might reveal) will not expose components of cognition, so the quantum-level semantics of DNA will never reveal the vitally important function of replication — a higher level meaning of DNA.

In sum, structure and meaning are relative terms and both can be differently manifest in various levels of explanation of complex systems. Moreover, explanation at one level will not necessarily provide useful information about other levels.

The trend in AI has been to build meaning onto a structural framework. Language processing systems, to continue with a familiar example, might be full of "meaning markers". Using this strategy progress was made towards the goal of getting words, and hence sentences, to carry meanings. Thus, in terms of the grammar rules used in the "recursion" chapter, this structural definition, slightly extended and with some attached "meaning markers", might be:

<sentence> is <noun phrase>$_{[Type1, Number1]}$

<verb phrase>$_{[Type1, Number1]}$

<noun phrase> is <det>$_{[Number1]}$ <noun>$_{[Type, Number1]}$

<det> is A$_{[singular]}$ or THE$_{[singular or plural]}$

<noun> is DOG$_{[animate, singular]}$ or DOGS$_{[animate, plural]}$

or $\text{CAR}_{[\text{inanimate, singular}]}$ or $\text{BAT}_{[\text{animate or inanimate, singular}]}$

or $\text{BATS}_{[\text{animate or inanimate, plural}]}$

The semantic markers are displayed as sub-scripts between square brackets. There are two types of semantic markers in this example: marker names, designated by an initial capital letter, such as "Number1"; and specific values for the markers, designated by bold lower case letters, such as "**singular**" or "**plural**". Thus during potential sentence analysis or generation, the marker "Number1" may by assigned the value "**singular**" or "**plural**". It is these grammar-defined marker relationships that are designed to add meaning to the sentences of the language (and to eliminate grammatical nonsense). How does it work?

The associated markers are carried from rule to rule, and checked for the obvious (but in this toy grammar unspecified) necessary restrictions. For example, in the <noun phrase> rule, the "Number1" marker value for <det> must be the same as the "Number1" value for <noun>. The result of this restriction is that a potential sentence starting with "A BATS…" would neither be generated nor accepted, because "A" carries the "**singular**" value as its marker, "BATS" carries the "**plural**" one, and the <noun phrase> rule requires that these two values of the "Number" marker are the same (denoted by a post-fixed digit, i.e., both grammatical constituents have "Number1" as their marker name).

But this sort of augmentation can only get the language processor so far. We already see further complexities in just the tiny bit of grammar given above. Thus the <det> word "THE" can be either **singular** or **plural**, so "THE BAT…" and "THE BATS…" are both acceptable, as they should be. But the complete sentence cannot be "THE BAT ARE FLYING". This is ungrammatical because the singular form of the noun phrase clashes with the plural form of the verb. So a loose restriction within the <noun phrase> rule must be rigidly enforced between the actual <noun> and the following <verb phrase>.

Meaning is added in the sense that the word "BAT" can be a baseball bat or a small flying mammal. In a sentence such as A BAT FLEW OVERHEAD, the marker value **"animate"** for the word "BAT" is likely to be selected as the best fit with "FLEW OVERHEAD". Arguably, this choice has begun to associate an appropriate meaning to the use of the word "BAT" in this sentence.

Initially, such peculiarities can be accommodated within the system, but the complexities of English soon defeat the enterprise. Consider, for example, the ramifications if the "THE BATSMAN HAD LOST HIS GRIP" follows "A BAT FLEW OVERHEAD". Over the years, far more imaginative and complex ideas than my simple attempt at a meaning-markers strategy have failed to deliver the goods. Following the AI tradition the initial project exhibits some interesting sentence analysis or generation capabilities; some erroneous behaviours become apparent; the system is modified and extended; new erroneous behaviours emerge; the project is further extended. This process, which results in ever-increasing system complexity, terminates when the ingenuity and programming skill of the researchers is eventually stymied by the difficulties of the fixing required.

The new (actually not so new, given the long-term quest for "semantic information retrieval" capabilities) challenge is for "semantics-based" searching of the Internet, i.e., to replace the current keyword-based retrieval with one based on "meaning". The current, spectacularly successful search engines do, of course, rely heavily on intelligent filtering of the retrieved items. It is, however, you and I that must provide this sifting process. Perhaps, one day it will be the computer. The term "semantic web"[5] is much bandied about, and various "better than Google" developments of search engines are "well underway".

The IT entrepreneur Stephen Wolfram, for example, is promoting "Wolfram|Alpha"[6] as a new "knowledge-based" question answering system. It is expected to be preferable to Google's list of possibilities that a user has to read in order to determine which are relevant. The initial launch in 2009 of this "Google killer" failed to live up to expectations, but has now been further expanded to about "10 trillion data points, making it, Wolfram claims, the largest integrated data set in the world" (p. 32). The upfront challenge for this system is for it to understand the questions that users pose. It clearly must tackle a far wider range of options than Watson, the *Jeopardy*-playing system, does, but Wolfram|Alpha's language-understanding capabilities only need encompass likely question-asking patterns. This may be the reason behind Wolfram's declaration that "the language understanding problem... it turns out... was a lot easier than we thought" (p. 34). He is referring only to this one small domain of natural language possibilities. Success in the Turing Test is still way over the horizon, despite declarations of "how surprisingly easy" natural-language understanding has turned out to be. The big breakthrough is still only in hopeware.

Ever since the 1960s when the computational models of AI researchers first began to exhibit interesting behaviours and appeared to be on the verge of really intelligent performances, the crossing of this Rubicon has continually been thwarted by failure to solve the so-called "frame problem." This obstacle, in terms of our current preoccupation with meanings, is manifest as a crucial necessity for global knowledge — information outside whatever framework we are using. Typically, the intelligent observer can "see" what extra information (external to the system as constructed, and hence "framed") must be tossed in to circumvent observed performance inadequacies.

But when the system's compass is extended to embrace the more global information (from say simple meaning makers for words, as illustrated above, to sentence contexts), the problem of meaning resurfaces further out — just like a mirage in the desert. Meanings will arise in the context of the full text, and new meanings beyond, to ultimately be determined by the beliefs of the writer, or reader. These last two perspectives on the meaning of a set of sentences may well be different, perhaps spectacularly different. For the most part, human language works quite well. From this viewpoint we conclude that sufficiently similar meanings tend to be extracted, otherwise our languages would not be effective as a means of communication.

Yet again we see a key aspect of intelligence that is founded on compromise — human language is certainly not a perfect means of communication, but it's good enough, most of the time. In an earlier chapter, we've also seen that elevation of our aspirations to the level of a super-AI will not eliminate this need for compromise. On the contrary, it is likely to exacerbate the problem of communication with humans.

In addition, targetted performance inadequacies of any specific system may well be eliminated by adding extensions and patches. But then others are likely to emerge, and yet others, which were around all along (but perhaps not initially a frontline concern), will still need addressing. The system, of course, gets larger, more complex and therefore more unwieldy with every such expansion of the "frame". The classic AI remedy is to fall back on special-purpose ad hoc procedures, called *heuristics*[7] but this only pushes the problem out beyond the current frame to the need for global information to choose the appropriate heuristic.

Frames tie up with modules — if a complex AI system is composed of maximally encapsulated modules, then the chances of managing it and maintaining control of it are similarly maximised. Ideally, the complexity of natural-language processing, for example, would be manifest as interacting modules of specialised linguistic competences[8] — the language and linguistic expectations for dealing with restaurants, for dealing with car journeys, etc. Each such domain of local linguistic competence could be captured in a template of specialised vocabulary and "plans" to structure expectations (such as, "ordering $\rightarrow$ eating $\rightarrow$ paying" in a restaurant). Each such module, or perhaps set of closely related modules, corresponds to the idea of a "frame", and hence which meanings are within its bounds, and which are not.

So, for example, the meaning of the question, "Do you want to see the menu?" is to be expected within a "restaurant frame", whereas "Do you know the theorem of Pythagoras?" will be outside this particular frame.

What about "Where are the toilets?" or "Can I sit by a window?", and countless other possibilities that might well arise in a restaurant scenario, but can they all be in the "restaurant frame"? And what about eateries where "order → pay → eat" is the order of events, or even fixed-price establishments where "pay → order → eat" is the sequence?

You and I can handle all this uncertainty more or less effortlessly (accepting the occasional hiccup). You have a complex set of expectations related to eating in restaurants but the idea that this set has a firm, or even firm-ish, boundary is hard to square with the apparent flexibility and seemingly boundless eating-out skills of you and me. Although simultaneously walking and chewing gum has been said to be a challenge for certain individuals, to call this restaurant competence "intelligent", or even "clever", is a bit over grandiose. Yet science has little idea of how we manage to negotiate dining out and the associated verbal communication even if our apparently smooth and seamless behaviour is really based on frame-to-frame transfers and interactions.

The connection between frames containing all the hopefully pertinent information (or at least an amount sufficient for intelligent behaviour) and informationally-encapsulated modules is straightforward. Consequently, the problem of the continual necessity for external (and therefore more global) information, casts this AI-system awkwardness as an old friend — the modularity-holism conundrum, the puzzle that undermines the basis of all the proposed cognitive architectures.

In Fodor's words, "The 'frame problem' is a name for one aspect of the question of how to reconcile a local notion of mental computation with the apparent holism of rational inference... the fact that information that is relevant to the optimal solution... can come from anywhere."[9] In a restaurant it is far from unusual to chat about topics other than food, menus, service and bills. Should you find yourself dining with a friend who has just been enthused by a geometric revelation, then "Do you know the theorem of Pythagoras?" may well crop up in the table talk. If you are to make sense of this and engage with your friend's new interest, then presumably a "geometry frame" must be used within the "restaurant frame".

A few pages earlier, Fodor makes the sweeping claim that "the failure of our AI is, in effect, the failure of the classical CTM [computational theory of mind] to perform well in practice". He sees a pattern in the failures of AI to produce successful simulations of routine common-sense cognitive competences[10]; in his words, it's "notorious, not to say scandalous".

AI researchers have long lamented the "frame problem". Wherever the lines are drawn for semantic knowledge, the neglected "meanings" from further out pop up to undermine the intelligent activity of the system. I think that the following quotation from Lawrence Durrell sort of sums it up:

> We get too certain of ourselves travelling backwards and forwards along the tramlines of empirical fact. Occasionally one gets hit softly on the head by a stray brick which has been launched from some other region.[11]

The primary stretch beyond reasonable literary licence here would be solved if "occasionally" was replaced with something like "pretty quickly". But by pre-dating the general acknowledgement of the frame problem (not to mention the utter unlikelihood of Durrell ever getting wind of it), it's not at all surprising that he didn't hit the frame problem right on the nose. Today, we have no similar excuse, although both Pinker and Plotkin[12] apparently give it a miss.

Notice also that the general strategy for reproducing human-language abilities with a computer system tends to be founded on the notion of a complete, initial syntactic analysis. It is, however, an empirical fact that a lot of human-language communication is manifest in terms of incomplete and ungrammatical sentences, and yet it works (most of the time). There have been some worthy attempts to allow semantic models of language to drive the production or analysis of language and so eschew complete, prior use of grammars[13], but the removal of the only (relatively) solid foundations does not facilitate the construction of a semantic edifice*.

An intriguing alternative, the conversational avatars, first George and now Cleverbot, have won a few Loebner prizes, among others, for the "most human" conversational system[14]. The unusual strategy employed to produce these chatbots makes no explicit attempt to tag meanings onto the words and phrases, nor to build an elaborate knowledge base or semantic network. Instead it relies on the necessary semantics emerging from the context of all the utterances that the avatar has experienced.

This strategy avoids virtually all grammatical and meaning analysis, and relies instead on the collection of things that people typically say in response to whatever utterance it is dealing with to generate its own response[15]. Two extreme views suggest themselves: either Cleverbot (if ultimately it does achieve the linguistic competence of a human) has learned an adequate unconscious semantics of human language (but probably not the more sophisticated semantics of consciously learned knowledge); or alternatively, Cleverbot's linguistic abilities are virtually meaning-free, in the style of a complex stimulus-response capability (which in turn suggests that development might stall at a threshold where stimulus-response is no longer enough to carry the implied semantics). But which of us has never engaged in a personal-meaning-free conversation when cornered by a voluble stranger at some social event? The ruse can work for a while but not for long

---

*How would the ultra-AI approach this further "oddity" of human language usage? Insist on total grammaticality in an effort to ensure "correct" meaning, or go with the flow of normal human communication and its inevitable mis-communications?

enough to pass the Turing Test we must presume, although successful maintenance of such a charade may also indicate intelligence.

The latter Cleverbot alternative is suggestive of a dismal failure with respect to the quest for language meaning, but is it? Ultimately, one can argue that meaning is to a large extent not resident within the system being interrogated. To an external observer, which is all we can realistically ever be with respect to an AI system (not to mention everyone else we encounter), the final semantics of the system's observed behaviours is in the mind of that observer. The semantic implications drawn are largely based on what the observer would mean if he or she performed the behaviour observed, mediated by the observer's beliefs about the system observed. Once more, effective communication between "systems" (perhaps a human and an AI system) is called into question.

From a semantics perspective, effective communication must rest on sufficiently similar "meaning models", or "value structures", within each system — ours is shaped by living with a human body in the world that our minds and bodies evolved within. What will be the basis for the AI-system's? Will we install a pre-computed replica of ours? Possibly, but that's one heck of a challenge on top of the already daunting task of replicating the basic functions of intelligence. One result of this strategy (even if we could hope to do it), would be that the AI system will be "living" a lie, and how do we ensure that it can keep this fiction up to date without a human body and human lifestyle?

Pushing further out to probe the origin of meanings, consider that the world is a dark and silent place crowded with a tumult of energy fluctuations. We humans live in a raging sea of waves (air-pressure waves, electro-magnetic waves, waves of molecular concentrations, etc.), but individuals of the human species are aware of only a very small part of this "blooming, buzzing confusion"[16]. We "see" a small part of the electromagnetic spectrum — the so-called visible spectrum — and we perceive a few specific sharp distinctions between very similar waves, e.g., the colours red and green. We sense other parts of this spectrum as, for example, heat. We "hear" a small range of air-pressure waves, and are oblivious to the potential sounds either side of the audible span.

The history of human evolution "explains" the specific and limited parts of the energy tumult that we happen to perceive. The structure of human reality is determined by the necessity to survive and reproduce. We are aware of, and perceive in specific ways, just those selections from the entirety of energy fluctuations that facilitate human survival and maximise the likelihood of reproduction.

The totality of energy fluctuations is in itself valueless, meaningless. Meaning and value, and hence emotional response, accrue only as a result of the function of a perceived stimulus in furthering the likelihood of survival and reproduction. Sugars taste good because they are important for survival. Our waste products smell bad because ingestion would be detrimental to survival. Orgasms feel particularly good only because of their crucial role in reproduction, and so on.

Bats, for example, are largely oblivious to the human visible spectrum, but hear (or perhaps "see") a world of obstacles and prey in terms of air-pressure waves. To us humans their meaningful stimuli are meaningless, indeed largely invisible.

What does this rationalisation of human emotions and the meanings of perceived stimuli have to contribute to the scientific understanding of intelligence? It raises serious doubts about the validity of the assumed separation between minds and brains (the CTM, you may recall). If the meaning and value of stimuli are epiphenomena that evolved, within the human brain-body, specifically to promote the human species, these characteristics *cannot* be *in* the stimuli, i.e., not in the world at large. Intelligence (insofar as it must be based on a complex value structure and associated emotional responses) cannot then be a computer program totally divorced from the details of the "machine" on which it runs. If so, could a computer program (even

357

if it incorporates a simulation of basic human values) ever be expected to pass the Turing Test?

From the CTM viewpoint, the complex value structure that could be an essential ingredient of intelligent behaviour may be dependent upon relatively few and readily reproducible aspects of human brains (and perhaps bodies). So the CTM in conjunction with a few, well-defined "machine" characteristics might constitute the best hope for the scientists who aspire to understand intelligence as an abstract concept, i.e., as a computer program tied to minimal machine requirements.

If we are content with an understanding of the basics of intelligence (and if such a thing is possible), perhaps we can largely ignore all the fictional (for the AI system) attributes that come with being a human rather than a computer simulation. But what would its semantic frame of reference be like? Not just like ours, for sure, but would it be startlingly different? It just might be. It almost certainly would be in some respects.

As simple a thing as an apple, for example, means so much to me and you (I assume because we lead similar lives in similar ways). Sure, both you and the AI system know it's a fruit that grows on apple trees, and that it's edible, etc. But what can the AI system ever know of its taste or texture, or the feeling of satisfaction when eating it assuages a hunger? You and I know all these things through experiences in the world with the bodies and consequent minds that evolution has foisted upon us, but the AI system either does not know them, or has had the fictional knowledge implanted. Neither alternative seems satisfactory.

The chase for the ultimate meanings, the value structure, of intelligent behaviour has a frustrating tendency to keep leading the hunter away over the horizon. As soon as a concerted effort to close in on it is instigated, it slips away again. This quest may only end when an AI system is out and about in the world. Only then will we stand a chance of finding out what it really means by observing what it does, and in what situations. My suspicion, though, is that sorting out the internal meanings generated by what we've constructed could be quite a challenge.

Endnotes

[1] This quotation from C. K. Ogden and I. A. Richards classic work *The Meaning of Meaning* (Harcourt, Brace & Co., 1938), pages 12–13. It continues "and the vast majority of those who have been concerned with it [i.e., language semantics] would have been right in their refusal to discuss it".

[2] From the conclusion (p. 233) of Robert K Lindsey's article on "Natural-Language Processing" (pages 217–233) in *Computers and Thought* (McGraw-Hill, 1963) an influential compendium edited by E. A. Feigenbaum and J. Feldman.

[3] *Semantic Information Processing* edited by Marvin Minsky (MIT Press, 1969).

[4] From an article entitled "A real science of mind" in the online *New York Times* on 19th December 2010 by Tyler Burge who criticises primarily MRI brain scanning as overhyping the possibilities for elucidating components of our cognitive architecture.

[5] A current hot topic is the so-called "semantic web" but this, as currently conceived, is but an elaborate scheme for semantic tagging of web documents (either by explicit semantic marking, or by use of new "web publication" languages). Its success will perhaps enhance the already amazing information-retrieval possibilities of the Internet, but it is not a move to "meaning-based" retrieval rather than the current keyword-based process.

[6] This write-up on Stephen Wolfram and his new "Google-killer" is to be found in Alex Bellos' article "Can this man topple Google?" in *The Guardian Weekend*, 12th February 2011, pages 32–36.

[7] The heyday of *heuristics* in AI systems is perhaps past. The name is derived from the Greek for "steersman". These special-purpose, handcrafted procedures could always be counted on to solve the immediate inadequacies of any system — all that was required was human creativity. The perceived need for many *heuristics* provided one persuasive argument for intelligence as a collection of many special techniques rather than the more scientifically satisfying basis of a few elegant general principles. Their value for surmounting awkward problems in AI systems (indeed all IT systems) has not diminished, as our later examination of the Watson system will reveal. The change has been that the word has lost currency as the worth of the strategy has been downgraded.

[8] Examples are Roger Schank's restaurant "scripts", and Yorick Wilks' "semantic templates", both described with examples, and fully referenced to original sources in my *New Guide to AI*.

[9] From Jerry Fodor's *The Mind Doesn't Work that Way* (MIT Press, 2000), p. 42.

[10] "We still don't have the fabled machine that can make breakfast without burning down the house; or one that can translate everyday English into everyday Italian; or one that can summarize texts; or even the one that can learn anything much other than statistical generalizations." p. 37 of Fodor's book, Endnote 9.

[11] From Lawrence Durrell's *Clea* (Faber and Faber, 1960) p. 205, the final volume of his four-volume (three of space and one of time) classic *The Alexandria Quartet*.

[12] Fodor notes disparagingly that neither Pinker's *How the Mind Works*, nor Plotkin's *Mind Evolution* find the frame problem sufficiently important to index it (see Endnote 9, p. 42).

[13] Yorick Wilks' "preference semantics" and Roger Schank's "scripts" and "plans" are just two examples from the heady days of AI research seeming to sweep all problems along before it.

[14] "Cleverbot wins Machine Intelligence Prize", *Cambridge, UK, 15th December 2010*. A special version of the Cleverbot application won the British Computer Society's Machine Intelligence Competition 2010, after taking part in a quick-fire Turing Test. Cleverbot was running with notably more power behind it than is possible for the online version, with 24 separate instances conferring on their answer. Ten volunteers talked for two minutes each using a plain text interface, and the whole of the event audience voted on how "human" each conversation appeared to be. Cleverbot achieved an average rating of 42.1% human!

[15] Cleverbot's designer Rollo Carpenter writes (personal communication, 2nd March 2011) that the chatbots' conversational capability is based upon something like "52.9 million rows [which] is the number of conversational pairs (what the chatbot said followed by human response) in the database, but that's a moving target since it grows by 60,000 to 100,000 a day. Of course it stores much more than just pairs, and refers to the whole context of conversations. There is some pre-processing of the data, and caching but mostly it is queried (in many ways) and *some* of the data is analysed fuzzily in "real time" every time. That's because almost no two contexts are ever the same".

[16] This famous phrase from the psychologist William James' classic *The Principles of Psychology* (Macmillan, 1890), vol 1, p. 488 refers to "The baby, assailed by eyes, ears, nose, skin, and entrails at once, feels it all as one great blooming, buzzing confusion". Evolution has focussed our perceptions on just those bits needed for survival, and the detailed nature of human intelligence must be key to the integrated filter we all possess.

⚔ Chapter 15 ⚔

Hopeware Science

"[As a consequence of their Fifth Generation computer project] by the beginning of the next decade, the Japanese plan to be well on their way to utilizing the amassed knowledge of human civilization."
1984 commentary designed to trigger a US strategy for AI[1]

"Consciousness in the sense of access is coming to be understood."
Steven Pinker, 1997[2]

"DeepQA is an architecture with an accompanying methodology, but it is not specific to the Jeopardy Challenge... We have begun adapting it to different business applications and additional exploratory challenge problems."
Designers of Watson AI system, 2010[3]

Recall the iCub robot from Chapter 1. It looks like a human child and even has some facial expressions. It speaks like a child, and it can do things that a human child can do. It can learn to recognise objects, and it can understand what it is told to do with these objects. In particular, we are shown that it can reach out and touch the objects it has just "learned". Other versions, we are told, learn to crawl, to shoot arrows at a target, and so on.

"You can almost see its artificial brain learning in front of your very eyes," the TV frontman, a scientist, points out. Sadly, rather like the persistent optical illusions in an earlier chapter, this is true — it is hard not to see AI at work, even though there is actually precious little objective evidence for it.

Why do AI presentations (even those at scientific meetings[4]), commentaries and media extravaganzas get away with this scientifically vacuous show-and-promise technique time after time? As you now know, I put the blame squarely on evolution aided and abetted by diverse sources of complexity.

Like all others throughout the history of AI projects, iCub is "hopeware". This is admitted when it's introduced as a project "in the early stages of development". It is time to delve into and try to explain this odd branch of science, this dubious escape hatch for the objective scientist whose bright idea is becoming bogged down in the swamp of ever more-challenging modelling complexities.

There's something of a clash between the two words of this chapter title — perhaps "Hopeware masquerading as Science" is preferable, but devoid of snap. "Jam Tomorrow" is a word couplet that promotes the flavour of this chapter, although I am of the persuasion that, in this case, tomorrow might one day arrive to reveal a sweet confection that resembles jam.

Any but the most casual browser of the preceding chapters will have clocked my use of the bizarre word "hopeware". Despite its failure to crash mainstream English, it has long been currency in the coffee-room argot of AI where its value resides in its similarity of both structure and function to those fundamental terms, hardware and software (with the brain completing the quartet as "wetware"). It is time to examine the "hopeware" phenomenon — why it occurs and how it works against our science.

Reliance on success just over the horizon is pretty secure. Horizons, just like tomorrows, may get progressively closer, but, in the normal course of events, they will never actually arrive to cause possible embarrassment. New dawns in AI, however, are almost as common as the real thing (see, for example, "I, algorithm: a new dawn for artificial intelligence", *The New Scientist*, 31st January 2011 by Anil Ananthaswamy) and, disappointingly, they break almost as quickly, or more often, fade away quietly into the harsh light of reality.

As a pertinent and well documented example, in 2001 the *The New Scientist* published a gushing article about a research lab in Israel where a computer was being taught to learn English.

> *As you enter the headquarters of Israeli company Artificial Intelligence, you can't escape the feeling that you're stepping into the secret lair of a James Bond villain. On the surface it might look like a regular luxury mansion, but below ground, in a bombproof bunker, they're plotting world domination.*
>
> *Jack Dunietz, founder and president of Artificial Intelligence (AI), prefers to call it a 'paradigm shift', but there's no mistaking his intention. 'If we're right, this is going to mean a profound change in our culture,' he says.*
>
> *Dunietz's secret weapon is a small infant called Hal who has never seen the light of day, spending his whole life locked in the basement. Sounds like a job for 007. But my mission was not to rescue Hal, it was to interrogate him and find out how much he knows.*[5]

It was a computer that was being taught English by being spoken to, as if it were a child. A few weeks later, *The New Scientist* published my letter pointing out that Hal (as the machine was hopefully named) could not be about to learn English by simply being talked to because "the power of this approach… is seriously undermined by the fact that we (the computer science and AI community) do not have any algorithms that exhibit learning behaviour with the power and flexibility of humans, even human babies (or perhaps, especially human babies). Our learning algorithms are crude, brittle, short-term and severely limited in scope. Machine learning, like every other aspect of AI, is a science of the future — and will remain so for the foreseeable future."[6] A contemporary BBC News report[7] stated, "According to Mr Dunietz, by the year 2010, computers like Hal will become part of our lives, talking naturally with humans."

Well, 2010 has come and gone, and I see no reason to revise my initial assessment. I see no computers with the capability for "talking naturally with humans", and none that remotely promise "a profound change in our culture" due to conversational ability. Notice also that, in sharp contrast to the 2001 avalanche of press releases promising Hal's breakthrough in intelligent communication, now that the delivery date is imminent (or perhaps gone, the original predictions varied) the news feeds are devoid of reports on Hal's progress. **By its fruits you will know it**, as Maxim 8 asserts. One suspects that life support has been quietly withdrawn. Young Hal is resting in peace, and we should not disturb him further.

It is perhaps the chatbot called Cleverbot[8] (introduced in the previous chapter) that comes closest to matching your chatting skills. In 2005 its forerunner, George, won the "most human" category in the Loebner Prize, an international competition for the best "talking" computer system that fails to pass the Turing Test (a Chapter 1 topic).

As mentioned previously, Cleverbot has learned its language abilities in an interesting way, similar to that overhyped by the Hal developers but without the necessity for non-existent machine-learning algorithms. Unlike the majority of chatbots, which are tightly bound by programmed rules and therefore distinctly limited in scope, the "Jabberwacky AI" software behind Cleverbot relies entirely on feedback — it makes fundamental use of contextual learning techniques to maintain its conversation.

In other words, the system stores everything it hears and responds by retrieving what is judged to be the most appropriate response from among all those it has stored as previous responses to similar questions. It eschews the detailed grammar and semantic analyses that most such systems employ. It relies instead almost entirely on efficient "memory" management and consequent information retrieval. So far, this unusual strategy (although arguably similar to human language learning) has produced a decent talking computer system, but can it attain the level of linguistic competence needed to pass the Turing Test? Or will its improving performance be swamped by the ever-growing memory burden, and its retrieval abilities similarly degraded by having so much information to search before it can reply? No one knows. Rollo Carpenter, Cleverbot's inventor, is optimistic that within the next ten years conversational AI will consistently be able to pass the rigorous questioning and analyses imposed during a formal Turing Test. I am less sanguine. Why?

Cleverbot is very good at maintaining what I would call a chit-chat conversation, so much so that Rollo Carpenter is constantly fielding accusations that the chatbot is a hoax (about one per day, he says)[9]. And this charge, in itself, is testimony to its communication abilities; it is daily passing a casual Turing Test (aided and abetted by Maxim 5). This is further evidence of the power of the genetic imperative that imputes intelligence from minimal provocation. From ELIZA[10] in the 1960s to Cleverbot in this new millennium a fluency in meaningless computer chatter has never failed to push our button.

However, Cleverbot cannot, and will never be able to, hold a long and detailed discussion on some topic that requires extensive "understanding", because no attempts have been made to build such "understandings" into it. One good reason might be that we have little or no idea how to do this, as the earlier chapters have made clear, I hope. While dealing with "hopes", one might counter this with the possibility that future generations of Cleverbot will build the necessary "understandings" piece by piece from their conversational experiences. Well, they just might, but that's well over the horizon once more.

An even more nebulous strand of hopeware is based on the published AI theories. The confident promise of a way forward is relatively easy to sustain when predicting from a textual theory. The tricky part comes when the fine words and confident statements have to be translated into hard code, a working computer system, a model. Hence my preoccupation with computer modelling (i.e., AI) as the necessary precursor of a sound scientific understanding of intelligence. A good and long-running example, aired in an earlier chapter, is Hebb's neuron-based "theory" of "reverberating cell assemblies", which is revived with stultifying regularity.

Introducing his "new theory" of intelligence, Hawkins asked, "Can we build intelligent machines?" and helpfully answered in 2004, "Yes. We can and we will. Over the next few decades, I see the capabilities of such machines evolving rapidly."[11] Close to the end of the first of these decades, rapid evolution of the intelligent capabilities of machines is proving hard to spot, especially if we discount the advances based primarily on faster and more powerful processors.

A popular[12] and similarly "promising" AI theory is the "society of mind" (SOM), which was expounded in text in 1985 but not programmed, by the AI leading light Marvin Minsky. It is an intriguing idea that intelligence is the result of cooperating "society" of experts. But just like Hebb's and Hawkins' "theories", so much remains unsaid that the researcher hoping to build, say, a SOM-theory model is left floundering amidst all the decisions he or she must make before a working model is obtained.

At best a working model with limited scope (if one is ever achieved) can offer nothing more than hope for what further development will deliver en route to intelligence. Discouragingly, the researcher can do no better than speculate on whether the difficulties encountered are due to "theoretical" failings, or poor (more or less) arbitrary decisions, or a combination of both.

As one AI researcher, who attempted to use the SOM theory to build a working model of human creativity, complained: "The biggest problem is to justify the claim that the way the model works is a reflection of the mechanisms embodied in the theory"[13] rather than all the somewhat arbitrary decisions needed to construct a working model.

Computer simulations of aspects of intelligence (as well as much more mundane functions) are detailed and complex systems. Even the system builder must entertain doubts as to whether the system does what it is observed to do for the reasons he or she believes it does (see Maxim 9). The further step of attributing system behaviour to theoretical elements can only be founded on inspiration, perspiration and hope — and so a whole new dimension of hopeware.[14]

One final way that hopes are maintained and prolonged is by evolving the system over time. The "Soar" system, another general architecture for supporting AI (a re-jigged successor to the GPS, the General Problem Solver, introduced below), provides us with an example.

When Soar was launched on the AI world in the late 1980s, the emphasis was on the (then) new buzz phrase, "learning at the knowledge level", and this learning was achieved as the central Soar innovation by "chunking":

> *Chunking creates new productions* [condition→action type rules, see Endnote 17 for an example], *or chunks, based on the result of goal-based problem solving. The actions of the chunk contain the results of the goal. The conditions of the chunk test those aspects of the pre-goal situation that were relevant to the generation of the results*[15].

Fast forward nearly quarter of a century and we find the Soar project still going. It has, we are told[16], been through eight major versions between 1982 and 2007. Very tellingly, in the course of these modifications, the original key elements such as "chunking" and "knowledge level learning" appear to have been dropped, or least dropped from the forefront of the modern descriptions. Production rules still provide the basis for representing knowledge in the system, and over the years "Soar proved to be a general and flexible architecture for research in cognitive modelling".

Although the "universal" learning mechanism of "chunking at the knowledge level" was the foundation of the architecture at its inception, the 2008 description introduces "substantial extensions to Soar, adding new learning mechanisms and long-term memories". No mention is made of how they relate to "chunking" or "the knowledge level", respectively. One suspects that they don't, and that there is little in common between Soar circa 1987 and Soar in the new millennium. Beyond a production-rule[17] basis, the only constant appears to be hope for the future — "we also expect that these changes ... will significantly expand the breadth of human behavior that can be modelled using Soar."[16]

A thoughtful and probing examination of AI published in 1985 predated the Soar announcements, and so John Haugeland had only its precursor, the immodestly named General Problem Solver (GPS), to examine. He commented that "GPS was a dream come false. Its ideal of generality rested on several unfulfilled assumptions... that all problems... are pretty much alike... and... that formulating a problem is the smaller job, compared to solving [it] once formulated."[18] It is fairly clear that Soar, as originally conceived, did little more than rename GPS's weaknesses. It is not at all clear that either of these two "unfulfilled assumptions" have been circumvented, or fulfilled, in its latest incarnation.

AI has long been known as the science of hopeware. Jaundiced observers, with whom I confess to have some affinity, have been pointing this out for decades. The other camp, the optimists, which must count the Singularity partisans as fully paid-up members, circumvent the embarrassing failures and quiet deaths by liberal use of riders for their pet AI projects further bolstered by imaginative resurrections.

The hopeware qualifications, which permit unchallengeable certainty here and now without actually delivering anything concrete, can be selected from the full array in *The Singularity is Near* (to pick a recent compendium at random) — "well underway" (pages 148, 270 omits "well", 293, 461 & 479); "already well down this path" (p. 483); "we are learning to understand" (p. 128); "a work in progress" (p. 148); "are starting to do this" (p. 151); "can be simplified once the operating principles of the brain are understood" (p. 153); "now approaching the knee of the curve" (p. 154); "capabilities that exist at least in an early stage today" (p. 166); "we are beginning to understand the regions of the brain" (p. 191); "once we fully master pattern-recognition paradigm" (p. 261); "once we understand the principles of intelligence" (pages 265 & 296); and "not far from realization" (p. 482). This is a far from exhaustive selection, but pick up any gushingly positive spin on AI, from any era, to find many more of the same.

It all started, as I mentioned with regard to Alan Turing in the opening chapter, at the first dawn of modern AI. Since then it has survived all AI crises, such as when infamously overhyped projects failed to deliver on the promises made on their behalf, and it is blossoming strongly in this new millennium.

Allen Newell and Herbert Simon formed a partnership that operated in the vanguard of AI for decades. They[19] were taken to task repeatedly over predictions made in 1957 about AI's anticipated successes which have not materialised. In the 1970s they still claimed that "they happen to be pretty good predictions" because "we just vastly underestimated two things: first, how little, how few man-years, would go into this; and second, how much very specific knowledge had to get poured into it".

At first sight I take this as a major cop-out, all failed predictions can be maintained "as pretty good predictions" with such caveats. However, as a believer in the possibility of AI, I have to give some credence to the assertion that progress in AI would have been more evident if more resources had gone into it. What makes me reluctant to acquiesce to Simon's excuse is not the dramatic failure to crack any of the basic problems (such as Simon's example of what is "knowledge", let alone how do we acquire, store and retrieve it), but the difficulty I have in pointing out *any* significant advances towards solving *any* of them.

Marvin Minsky, already introduced as a leading AI researcher, writing in the *Science Journal,* described in 1968 the somewhat rudimentary state of grasping robots "soon" to be much better; and the primitive sensing of computer eyes that "soon will rival man's analysis of his environment"[20]. Do I need to point out that in this bright new millennium, there are no computer eyes that "rival man's analysis of his environment"?[21]

From Turing in 1950, through Minsky and other principal advocates of the AI spring and summer, to today's incautious researchers egged on by temptations of celebrity (and consequently improved access to research funding[22]), the firm expectation that success is just a project away never flags.

Hopeware, as they say, springs eternal... as well as universal.

And not just in individual research projects. In 1981 the Japanese Government frightened both Western Europe and the USA with their launch of the Fifth Generation project. This grand strategy for AI in Japan was announced, and promised a range of AI wonders within a decade. Massively parallel computers programmed with logic-based programming languages[23] using large databases of knowledge were going to crack machine translation, intelligent reasoning, etc.

Despite all the disclaimers to the contrary from the various vested interests (for in addition to the huge Japanese commitment, governments in Western Europe and in the USA[24] had been goaded into launching competitive AI projects), the central AI targets were never achieved. There were, of course, spin-off successes but with huge amounts of money spent, and years of dedicated research, it could hardly have turned out otherwise.

It's almost as if the USA's Moon landing project failed to land a man on the Moon or even get anywhere near to it, but was abandoned and hailed as a success because the heat resistant tiles developed had found useful application in industry. In the world of science, it's only in AI that failures are not even setbacks but are redirected successes.

As the US commentators[1] (hopefully) put it in 1984, "The difficult programming languages, the struggles to make different programs compatible, the problems of putting human knowledge into machine form — are to disappear, eliminated in the new Japanese Fifth Generation of computers." Alas, it did not turn out so.

Religion is that other field of human endeavour notorious for casting its predictions way over every horizon potentially accessible to the living. From time to time over-enthusiastic groups, failing to grasp the key idea, have predicted a new dawn for their followers on a specific date, and so suffered the embarrassment of explaining away a wrong "divine message".

In the summer of 2011, a misplaced "end of time" prediction was made by one Harold Egbert Camping[25], an American Christian radio broadcaster. He is president of Family Radio, a California-based radio station group that spans more than 150 markets in the United States. His recent "end times" prediction was that on 21st May 2011 Jesus would return, the righteous would fly up to heaven, and that there would follow five months of fire, brimstone and plagues, with millions of people dying each day, culminating on 21st October 2011 with the end of the world.

His widely-reported prediction has passed without the predicted incidents occurring. In the inter-apocalyptic lacuna, Camping revised his belief — a "spiritual" judgement had occurred on 21st May, and that the physical Rapture would occur on 21st October 2011, simultaneous with the destruction of the universe by God. Happily, this prediction was no more accurate than his previous one.

AI seers, who cannot rely on pure faith, have to be a bit more cagey, and the more mainstream Christian view is also a good deal more savvy: Jesus Christ taught that no man knows the day or the hour of the Lord's return — that's pretty much my view on the (first) coming of AI.

Way back in 16th century Florence, the established Church was less forgiving. The monk Savonarola may well have avoided the ignominy (not to mention pain) of going up in smoke if he had just resisted promising things in this world, and confined his promises to the next in the time-honoured tradition of his calling. With a far less draconian fate awaiting the similarly demonstrably failing predictions about AI, the pressure to cast them further ahead, but still well within the bounds of this world, cannot command the same intensity.

Why is the scientific quest to understand intelligence so beleaguered by over-optimistic expectations? It is true that every scientist desires to "see" what's beyond any current achievement, but in AI especially, minimal current achievement is typically swamped by grand expectations of what might come next. Every step is a virtual breakthrough.

In this penultimate chapter, the various threads — summarised in the ten maxims — can be drawn together to explain the hopeware scourge indigenous to the science of AI, in particular. It is an unhealthy manifestation that hampers true scientific enquiry because it records spurious, easy successes as scientific milestones that consequently draw in expertise and funding that might otherwise be used to make some headway in tackling the fundamental problems.

We have the interaction of a couple of major technical difficulties and some unavoidable truths to explain the plague of hopeware. First, what exactly is this "hopeware"?

> *Carefully choreographed demonstration systems that appear to exhibit (some aspect of) human reasoning capabilities, and thus promise a major breakthrough "soon".*

Quite unbelievably as we have seen, there is even hopeware that takes a further step away from reality — the "system" that is expected to do amazing things "soon" is sometimes no more than a collection of ideas rather than a concrete, working program.

The major technological difficulty — **no significant computer program is completely understood** — will obstruct any attempt by observers to determine with certainty how exactly the system does what it is observed to do. It also undermines any effort to predict what else it can do, and cannot do. Part of the explanation is the mystery, and hence awe, of programming. If a program exhibits some intelligent behaviours, what else might it be capable of, especially if just souped up a bit? No one can say for sure. Indeed, even hours spent untangling the intricate implications of the program details will deliver nothing more certain than a basis for conjecture. So no one can clearly refute any expectations that are articulated. This point gains credence because it is only the limitations of the human programmers that limit what their programs can achieve[26].

This difficulty is barely countenanced anyway because the genetic imperative — **if it's like me, it's intelligent** — will have triggered your conclusion at first exposure (and so obviate the technically demanding, and error-prone, effort of looking into underlying mechanisms).

All IT systems will defeat attempts to develop a secure and complete understanding of the reasons for their observed behaviours. In the case of systems that strive to behave intelligently the situation is worse — it is not just completeness that eludes us, it is inability to extract any manageable theory. Why should this be? Because of the major technical difficulty with our subject of study: **the mind exhibits no joints for science to carve at**. In other words, there are no small-enough-to-manage chunks to break off, demonstrate and explain. All or nothing are then the depressing choices, which pretty much leaves the scientist floundering whichever option is selected.

In an AI system the tricky nature of what is being modelled, i.e., intelligence, further reinforces the impossibility of extracting a full understanding. "Intelligent" versus "stupid" or "insightful" or "creative" or "almost-intelligent", are descriptors of system responses that can be impossible to clearly separate when the goal is AI. This uncertainty severely weakens a major plank of the framework that the IT-system developer expects to rely on, i.e., the ability to distinguish clearly between correct and incorrect system behaviours. **Intelligence admits no simple tests** (Maxim 2).

Our sad fact — **failure drives science forward** — tells us that successful demonstrations can, at best, only be treading water against the flow of the current of progress in science[27]. Recalling the Eleusis card game experiments in Chapter 2, scientific progress is seen as gaining the knowledge that a potential theory is false. The body of scientific knowledge, which western society is so rightly proud of, does seem to take a knock when viewed as the collection of theories for which we have no hard evidence that they are false. But that's what it is.

Aspirations for AI are reminiscent of the plans to land a man on the Sun once we fully master the principles of heat-management, work on which is well underway such that the necessary capabilities exist (albeit in an early stage) today. Can a Sun landing be far off?[28] Of course I exaggerate, but this is the nature of too much AI reporting, and not only from the media who are naturally more interested in startling headlines than boring detail. Sadly, proper science is founded on detail, often demanding and sometimes perhaps even boring, especially to the outside observer.

From the robot Shakey, trundling around Stanford University's corridors of AI power in the '70s, and Edinburgh University's Freddy struggling to identify a cup even earlier, this ploy has never failed. Most recently we've seen various TV channels confronting more human-looking robots: iCub already mentioned, and Luc Steels' communicating robots[29]. The game has been upped in visual presentation standards which can only reinforce the illusion.

True to form the cute little iCub did a couple of tricks — it "learned" two named objects and then reached out to touch the chosen one. Use of the word "learn" implies so much for you and I, but what did it amount to for the iCub? How many objects can it "learn"? What are the limits of its "learning"? Does it, for example, know the colour of the purple car? Can it even interpret this question? What are the limitations on its visual-recognition ability which must fall far short of yours and mine? Does it remember anything from one demonstration to the next? And so on, the pertinent questions are legion, and some of them we explored in earlier chapters, but none are ever asked in public.

Our intelligence detectors, which appear to be firmly wired in by evolution, are shamelessly triggered time and time again. Hopeware in AI has flourished on this "genetic weakness".

AI is the science of the future — and always will be. Something similar was once said smugly of Brazil being the country of the future. Now with year on year growth, at a level envied by both the USA and Western Europe, the quip does not look quite so smart, and so it may just turn out with AI.

Endnotes

[1] From a long article urging a US response to the Japanese Fifth Generation project, "The Fifth Generation: Japan's computer challenge to the world", Edward Feigenbaum & Pamela McCorduck. *Creative Computing*, vol. 10, no. 8, August 1984, p. 103. The authors are a prominent AI researcher, and an AI journalist and author (see Endnote 19) respectively.

[2] Steven Pinker's *How the Mind Works* (Penguin, 1997), p. 145.

[3] "Building Watson: an overview of the DeepQA Project" by David Ferrucci, Eric Brown, Jennifer Chu-Carroll, James Fan, David Gondek, Aditya A. Kalyanpur, Adam Lally, J. William Murdock, Eric Nyberg, John Prager, Nico Schlaefer, and Chris Welty in the *AI Magazine*, Fall 2010, vol. 31, no. 3. pages 59–79. Watson the AI "breakthrough" of 2011 is given full consideration in the next chapter.

[4] At such meetings the theories that underpin the AI system's demonstrations are, of course, discussed. This effort to elucidate the scientific basis is always corrupted because the only real basis for understanding the full scope of what the robot can do is the computer program that controls it. Many thousands of lines of program code cannot be presented as any sort of explanation. So a summary of the salient aspects of the program are presented. Quite apart from wholly unconscious construction of a "sympathetic" summary by the project scientists, it is impossible to avoid the use of words like "learn", "knowledge", "know", etc., which tend to trigger far more grandiose meanings in the minds of the listeners than the actual project warrants. This

long-acknowledged problem was tackled years ago under the rubric "rational recon-structions". The idea was that a complete and accurate summary ought to be repro-grammable by someone else to produce a system that would replicate the original behaviours, and hence substantiate the summary theoretical claims. The few attempts were not successful. See the articles in Chapter 7 (pages 235–265) in *The Foundations of AI*, edited by D. Partridge & Y. Wilks (Cambridge University Press,1990).

[5] The article entitled "Look who's talking" was published in the *The New Scientist*, 11th August 2001, issue 2303.

[6] My letter was published by *The New Scientist* on 1st September 2001, issue 2306, entitled "Laptop Lingo".

[7] Wednesday, 12th September, 2001, 10:37 GMT 11:37 UK BBC News "Computer babbles like a baby" by Joanna Chen in Tel Aviv.

[8] Rollo Carpenter, AI programmer and managing director of Icogno Ltd, and Tim Child, a pioneer of animated virtual beings in entertainment and the founder of Televirtual Ltd, developed the talking chatbot, George, and latterly, Cleverbot. These are a practical synergy of the conversational AI (or "chatbot") and an animated avatar (a virtual representation of an individual). Cleverbot can understand questions and speak responses. However, unlike his peers or his predecessors, Cleverbot's AI learns emotions and employs gesture and expression as well as language, allowing him to contextualise his responses in forms that are tantalisingly close to human and making him an entertainer as well as a communicator.

[9] In a personal communication on 1st March 2011, Rollo Carpenter wrote, "Many people tell me that Cleverbot is a hoax — that it connects live people together — but it is not true... Cleverbot replies come from a software program — only, ever, always. It learns what people say, in context, when those things are said. It imitates those people later — years later, even. So it can say exactly what you'd expect real people to say. You should NOT believe what it says — it might even tell you it's fake!"

[10] ELIZA was the creation of Joseph Weizenbaum. As a crude model of non-directive therapy, it worked surprisingly well (for short periods) by echoing back human state-ments (and throwing in bland requests when stuck, e.g., "Tell me more about your mother"). As a result there is a wealth of apocryphal tales of unsuspecting humans believing that they were taking to another person. As suggestions grew that ELIZA might really be usable in a therapeutic role, Weizenbaum felt compelled to publicise his opposition to this gross over-estimation of his creation. To read of his dissatisfac-tion, as well as some prescient musings on the dangers of our future reliance on com-puter systems, see his book *Computer Power and Human Reason* (Freeman, 1976).

[11] From p. 7 of Jeff Hawkins' book (with Sandra Blakeslee) *On Intelligence* (Henry Holt, 2004), expounding a cortex-based theory of intelligence. The theory, however, is very far from a complete and precisely formulated description; it is more the begin-nings of a newish theory.

[12] Expounded in Marvin Minsky's book *The Society of Mind* (Picador, 1985), and repeatedly and hopefully trotted out ever since, e.g., Steven Johnson's *Emergence*

(Penguin, 2001), p. 65, and Ray Kurzweil's *The Singularity is Near* (Duckworth, 2005), p. 289.

[13] This is said on p. 27 of *Computers and Creativity* (Intellect: Bristol, 1994) by D. Partridge and J. Rowe, and although a jointly authored book, all of the detailed modelling work (as well as this particular quotation) were extracted from the PhD dissertation of Jon Rowe.

[14] This particular difficulty with AI systems was aired earlier (Chapter 8) and has been treated at some length by a variety of contributors in *The Foundations of AI: a Sourcebook* (Cambridge University Press, 1990), edited by D. Partridge and Y. Wilks.

[15] A quotation from P. S. Rosenbloom, J. E. Laird & A. Newell's 1987 paper entitled "Knowledge level learning in Soar" and to be found in the Proceedings of the AAAI Conference, AAAI 87, pages 499–504.

[16] "Extending the Soar Cognitive Architecture" by John E. Laird in 2008, Artificial General Intelligence Conference, Memphis, TN. The brief history as well as the quotation are from the "Introduction" to this paper.

[17] A production rule is a rule of the form CONDITION→ACTION; in other words, if some CONDITION is found to be true then the associated ACTION is initiated. In an earlier chapter an example of a production rule was suggested as a possible implementation of iCub's unwavering delight every time it "learns" a new object; the proposed production rule was:

IF OBJECT "X" LEARNED → **SAY** "I like the X very much."where "X" is replaced with the new object's name as supplied by the iCub handler.

[18] The criticism is from p. 183 of John Haugeland's book entitled *Artificial Intelligence, the very idea*, (MIT press, 1985).

[19] Pamela McCorduck in her book entitled *Machines Who Think* (Freeman, 1979) reports (pages 188–189) on conversations with all the AI luminaries, and states that although the predictions and the subsequent apologia were both under Simon's name, they were the work of the two men in partnership.

[20] Another snippet from Hubert Dreyfus' book *What Computers Can't Do* (Harper & Row: New York, 1972). Minsky's article was originally published in the prestigious *Science Journal* in October 1968, p. 3, entitled "Machines are More Than They Seem".

[21] There are "seeing machines" (often using non-human technologies such as laser range finding) that outperform human seeing capabilities in all sorts of specific ways, but then vultures, for example, have always had the jump on us for distance vision. In terms of general "scene understanding" — i.e., looking around and seeing trees, cows, clouds, etc. — there is nothing "mechanical" that comes close to your amazing talent.

[22] This cynical aside may be readily dismissed by the science-funding organisations that typically operate strictly objective procedures for determining which proposed projects get funding, and which do not. Peer review (often with anonymity in both directions) is the usual basis for objective evaluation. But who are these "peers"? To

be qualified to evaluate they must in effect be either collaborators or competitors (both of whom will have enough detailed knowledge to see through the supposed anonymity) and so exhibit natural tendencies to gush in support or be brutally critical, respectively. These evaluations are then considered as a whole by a committee of the great and the good. Everyone participating in this total process is a busy and highly-qualified scientist, yet they all must give their time and effort for little or no reward (except that of being on the inside track with respect to funding decisions). So quite apart from vested interest in what gets funded and what does not, there is considerable pressure to deal with everything as quickly and efficiently as possible, and at all stages celebrity counts. Show business has no exclusivity claim to the value of "who you know" trumping "what you know".

[23] The European, if not British, logic-based programming language PROLOG was the best candidate as (or at least a basis for) Japan's chosen programming strategy. But PROLOG too has faded, for although it opened up really new programming possibilities, it similarly introduced new problems.

[24] See, for example, "The Fifth Generation: Japan's computer challenge to the world," Edward Feigenbaum & Pamela McCorduck. *Creative Computing*, vol. 10, no. 8, August 1984, p. 103.

[25] See, for example, Wikipedia entry for Harold Camping (accessed 10-6-2011).

[26] This "technical difficulty" is argued at length with copious examples and an analysis of the contributory factors from programming technology in *The Seductive Computer: Why IT systems Always Fail*, published by Springer in 2011.

[27] The history of science is littered with successful experiments that are rightly hailed as breakthroughs in our knowledge. How can this be? A pivotal difference is captured in the common quip: scientists try to understand the world, AI modellers try to create it. Classical science is primarily analytic and must deal with the world as it is, whereas the science of AI is mostly synthetic and enjoys the freedom to construct models that the empirical world barely constrains. Thus the classical scientist can boost support for a theory by demonstrating a novel application in the world as it is. The AI scientist may be demonstrating model-building skills rather than theoretical validity — distinguishing between these two extremes is tricky.

[28] A landing at night, I should add, is expected to contribute little to the possibility of a positive outcome.

[29] In the *The Hunt for AI*, broadcast in the UK on 3rd April 2012 at 21:00 on BBC 2, what we see is two robots learning to associate words with actions (such as raising an arm). Is this anything more than a naming of simple and carefully choreographed actions? It is no more than teaching a dog to associate "sit" with sitting on its haunches, and to associate "beg" with sitting up, etc. The arbitrary action-naming we see amounts to development of a "private language", but one devoid of virtually all elements of human language. However, we do see learning between robots (and there is no such learning between dogs). In addition, each robot uses vision to

identify the various actions, but what's the scope of recognisable actions, and what are the restrictions on the visual recognition process, e.g., will the action "left arm raised" be identified from behind the robot? Does the language-learning go beyond action-name associations? We are never told. When questioned about his achievement, Luc Steels pointed out that just getting a bipedal robot to stand is a huge challenge. Bipedal standing is a difficult problem of motor control, feedback sensing and balance. It is also almost certainly the case that human intelligence is skewed towards bipedal locomotion, and would be somewhat different if humans had three legs or used all four limbs to gallop about. However, is there any good reason to suppose that the development of a bipedal robot will make any positive contribution to the development of the cognitive processes central to human intelligence?

ଓଃ Chapter 16 ଔ

The Glass Half Full

"Consider that every intelligent creature in the universe bets its life
every day on inference processes that lack
performance guarantees."

R. Davis, 1991[1]

"Q: How significant is this match [between the Watson computer
and two humans] in the quest for artificial intelligence?
A: It'll be seen as a major threshold. The key to human intelligence is
really mastering the subtleties of human language, things like puns and
jokes and metaphors. And if you look at the queries in Jeopardy you
see they're quite complex and subtle, and exactly what's being talked
about is not so clear. Watson appears to be able to get it very well,
as well as the best players."

Ray Kurzweil, 2011[2]

Now is the time for truth, and reconciliation of the various points argued. A close look at Watson — the latest AI success — illustrates the differences between what has been achieved and what has been presented. This examination can be further used to give substance to many of the more nebulous claims about the diversionary pseudo-science engendered by AI projects in comparison to the real science that such projects embody.

Several positive proposals are developed, which may go some way towards countering the welter of critical comment contained in the earlier chapters. I am not an AI refusnik. I do deny that it exists. I also deny that it is just over the horizon. But I do not deny that it will some day be achieved, certainly a mechanisation and hence a firm basis for a scientific understanding of basic intelligence (if not the full-blown human version).

The ten maxims are presented as a fully integrated framework, a farrago of interrelated constraints with which the scientist must grapple when seeking an explanation of what makes us clever. Acknowledgement of these explicit difficulties, as well as the many subsidiary problems, is a necessary precursor to significant progress towards a scientific understanding of what makes us clever.

But, to begin, let's be brutally truthful about our early endeavours to extract the principles that underlie intelligence from close observation of functioning brains — so-called reverse engineering.

There is no known informationally-encapsulated brain module such as we explored in the opening chapters; and there are no such light-touch nanobot scanners. So why did I lead you through the elaborate charade? Let's put it in terms of our newly exercised scientific acumen: we tested the hypothesis that reverse engineering can succeed, i.e., that the proposed method can extract the underlying principles of brain operation and from them construct a computer system that is intelligent. This is the basic hypothesis underlying the proposed route to the Singularity.

By thoroughly exploring this hypothesis in the context of an ideal state of affairs, we could expose the futility of this superficially plausible route to understanding intelligence. The structure-to-meaning problem, the feature-salience-with-single-examples problem, the unmanageability-of-exponentially-many-possibilities problem, and the finding-joints problem all surfaced in terms of our specific example. In addition, this extended exercise gave us a first taste of holistic processing, a lead in for a subsequent detailed examination of this supposedly mysterious, but potentially key, phenomenon.

It was thus made clear that even in the best situation imaginable (carved-out brain modules and nanobot scanners), brain study alone is not going to deliver the answers sought. Something more, or something different, is needed. Recall that even the benefit of insights into the cognitive-level meaning of the module did precious little towards bridging the abyss between structure and meaning — the sought-after principles underlying the cognitive-level processing remained a mystery.

Early in 2011 the AI sensation was IBM's computer system called Watson. It hit the headlines by winning a US TV rapid-response question-answering game called *Jeopardy*. The AI system (which might well merit this label under my rules because it does make limited use of some learning technologies), trounced the two top human players by responding more quickly and correctly to a series of questions posed in English[3]. So is this the AI breakthrough so confidently, yet so patiently, awaited in some quarters? Is this the basis for a significant step forward in the scientists' understanding of intelligence?

The answer is "No" — or that's my answer (to confess to a degree of subjectivity here). Why? It is progress in AI, possibly significant progress — only time will tell. The basic system may pave the way for new and different applications in new areas, or it may contribute little more than some do's and don'ts of large-scale AI-system design. Given the single-shot-success history of landmark AI projects, the evidence we have (and will review below) of Watson's design principles offers precious little indication that this is something refreshingly different, a project that breaks significant new ground on the path to AI.

Watson is not, and was not designed to be, a generally intelligent system. It was trained automatically (and in this sense, it "learned") to answer *Jeopardy*-type queries. Despite the fact that the UK's BBC television programme *The Hunt for AI*[4] told us that "Watson understands the subtleties and nuances of English", it has no such competence. You cannot hold any sort of conversation with Watson. It is not even a general question-answering system. More limited still, you cannot expect an answer to a question that falls outside the tight remit of the *Jeopardy* paradigm. It is a finely tuned and optimised *Jeopardy*-playing system (and only that when two "awkward" categories of question are excluded[5], which the BBC failed to mention). Its impressive performance relies to some (impossible to quantify but significant) extent upon thorough prior analysis of *Jeopardy* questions and answers.

Rather like the industrial robot on a car manufacturer's assembly line, you may well be amazed and impressed by the quick, slick, high-quality welding you see it doing — smoother, faster and better than any expert human welder. But this level of automated performance can only be achieved on a precise welding task within a set-up that must be tightly controlled. Once the task is changed ever so slightly, or the car parts are not precisely in place, the fragility of the impressive performance is revealed. So it is with Watson. But when the performance involves language rather than, say, welding, our judgement is a good deal less rational.

The design decisions behind Watson's impressive performance are tightly tailored to the task at hand — playing *Jeopardy*. There is always something of a choice between a design that will work pretty well on a general component of intelligence, such as question analysis, and a finely tuned version (or even a totally different version) that will deal exceptionally well with only the few particular question types encountered in the task at hand. The former strategy holds some promise of progress towards general AI, the latter probably does not. But plumping for the latter course of action is always going to deliver the more impressive demonstration.

Watson does work from queries in English, and it produces many correct answers in English. Does that mean that natural-language analysis, semantic interpretation, and sentence generation have been cracked? No. It means that the very limited query types used in *Jeopardy*, their meaning in terms of database look-up for the "answering" facts, and the simple re-packaging as a *Jeopardy* answer have been cracked. This is no trivial endeavour, but neither is it close to the average human competence in language usage, although it is probably superior in factual knowledge retrieval skills (of the kind demanded to play *Jeopardy* successfully).

The designers state that "the over-arching principles in DeepQA [the core of Watson] are massive parallelism, many experts, pervasive confidence estimation, and integration of shallow and deep knowledge"[6]. And that all sounds good, but what's underneath the fine phrases?

Following in the tradition of IBM's other "big non-breakthrough" in AI, the Deep Blue computer system that beat world chess champion, Garry Kasparov, back in 1997[7], Watson's fundamental strategy is brute force — try everything that might work (in terms of extracting the meaning of the question from the bare words, and in terms of various strategies for finding answers to each potential meaning). A vitally important and, to judge by results, spectacularly successful addition was procedures to assess the confidence that Watson associates with each of its answering strategies. Apparently, a confidence in the correctness of each of its very many threads of alternative answering schemes is computed as they progress from text analysis through database search to answer generation. Then the "most confident" answer is the one finally presented.

Perhaps something like this:

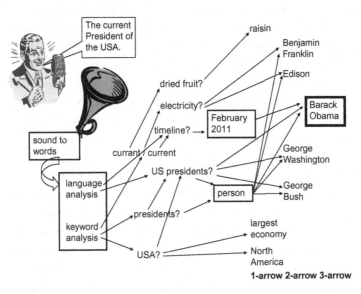

385

In this wholly conjured-up diagram of an imagined Watson strategy, the arrows illustrate supporting information and the *number of input arrows* gives the confidence to be placed in each particular stage in the overall processing. So "Barack Obama" (uniquely with "3-arrow" confidence) emerges as the most confident answer, followed (with equal "2-arrow" confidence) by "Benjamin Franklin", "Edison", "George Washington" and "George Bush".

"US presidents" is the highest confidence interpretation of the query. In the "Barack Obama" answer, it becomes coupled with the "time-line" interpretation of the spoken word "current" which gets a higher confidence rating than its two alternatives, "electricity" and "dried fruit" because these two interpretations get no support from the "language analysis" component (i.e., the timeline interpretation of "current" is the only one that makes sense in the context of the complete query).

Because of the massive proliferation of possible answer-generating strategies (and despite the undoubted use of the fastest processors that IBM could lay its hands on), it initially took about two hours to answer a typical question. At that point the system developers called in the parallelisation experts to configure a massively parallelised system. The idea behind this move is to use many computers, and give each computer one of the possible question-answering schemes to work on. In this way all of the many identified possibilities can be explored at the same time which reduces the computation time to that of the longest single possibility (or even the shortest that produces an acceptably high confidence in its answer).

By adding many more computers to the task (and further refinement no doubt of the built-in *Jeopardy* characteristics), the two hour response time was reduced to a few seconds. Simple (perhaps simplistic) mathematics, dividing two hours by, say, six seconds (the maximum acceptable response time), tells us that the parallelisation must have amounted to about 1,000 processors[8]. Although a daunting prospect for the average AI researchers, it was probably nothing much for a company with the resources of IBM and a massive public-relations coup in the offing.

However, chucking more computers at the problem is never this simple. Actually getting your hands on the machines is the (relatively) easy part. The great challenge here, and one that IBM's specialist team clearly rose to, is integrating all these processors. Although it may well have been the case that each processor, once allocated a question-answering strategy, could proceed more or less autonomously and get on with generating its answer with a confidence rating. Then only at the end, when all (or enough) of the independent parallel processors have produced an answer, is this collection of alternative answers and associated confidence ratings compared to select the winner. (In the live show on the Internet, Watson actually displayed its confidence ratings in its top three candidates for its eventual answer.) But regardless of the scope for simplifying the parallel processing, the effective integration of so many processors was no mean feat.

The confidence ratings, which appear to be crucial to Watson's success, are particularly interesting because they have been such a bugbear in cognitive science and AI research. Effective confidence estimation could well be one of Watson's major contributions to the science of understanding intelligence.

As an intelligent processor, you must sometimes be aware of a conscious effort to assess the relative likelihood that different answers are correct, or that strategies for obtaining an answer will succeed. Although we must exercise our customary caution in reasoning from our conscious activity to what the brain may be up to unbeknownst to us, this comparison is useful. Intelligent behaviour always involves some degree of assessing the merits of one strategy in comparison to those of an alternative. The assignment of confidence ratings to reasoning strategies, and hence to the possible answers that each might deliver, is managed somehow within our cognitive architecture. Consciously, we do it on a small scale; it may be that subconsciously our brains do it on a massive scale just like Watson (after all, brain scans always reveal massive parallel activity in our brains). Perhaps, but how? It just might be something like "amount of supporting evidence" that was crudely illustrated with the "number of arrows" example given above.

But like so many other bright and promising ideas in this domain, scientists can conjure up examples of how it might work quite well in pre-determined situations. But when they have to produce a generalised and precisely-defined procedure (i.e., a computer algorithm) that will work similarly effectively in a wide range of unforeseen situations — as intelligence must — they are likely to fail dismally.

Computer technologists designing AI systems have long favoured logic-based reasoning: firstly, well it's logically sound which sits well with scientific thinking (even if not so well with everyday human intelligence); and secondly, it's well-defined and can provide guarantees. But formal logic is based on just two assessments: true or false. As soon as uncertainty enters, which is pretty much immediately in any realistic context, there is a need to deal with assessments such as "maybe true", "probably true", "almost certainly false", "usually false", and so on. Then the logic is left floundering in the swamp of alternative schemes for quantifying confidence assessments and combining them as the reasoning process progresses[9].

Again, judging by results, Watson seems to embody a good solution. Although no doubt tricky and solved by the imaginative use of a variety of techniques, the question-answer (QA) context is easier than many. Why? There is no long chain of reasoning along which confidence levels must be managed (i.e., raised and lowered appropriately as the reasoning progresses).

In the typical chains of reasoning that appear to be involved in intelligent decision making, the primary stumbling blocks have been:

1. how to quantify uncertainty assessments; and

2. how to combine uncertain assessments to produce an uncertainty estimate for the next step in the reasoning process.

The former problem is viewed as putting numerical values on the casual assessments that you and I make. So if say "totally confident" is 1, and "absolutely clueless" is 0, what value is to be put on "not-very confident" or "quite confident"?

The second stumbling block adds the issue of how to combine less-than-totally-confident assessments. Let's say you've lost your credit card. If you're "fairly confident" (say, 0.7 confident) that you last used it yesterday in the local supermarket, and think that it's quite possible that you may have not removed it from the machine (say, 0.3 confident), then how confident should you be that the supermarket is the most likely place to enquire about it? Boiled down to essentials what is the confidence rating produced by the combination of two individual confidences 0.7 and 0.3?

Both of the above-listed issues are wide open.

In this QA system tackled by Watson, the query is broken up into key words that trigger searches and produce potential answers. Then confidence in these answers can be founded upon elaborate comparisons between the question and the potential answers. A variety of procedures compute matching scores along various dimensions — is the answer, perhaps a person's name, an appropriate sort of answer for this question? Does the named person's timeline fit that of the question? And so on. The designers of Watson needed to devise the various comparison procedures, and they also needed to merge the various results.

The Watson researchers have clearly come up with a number of innovative procedures for all sorts of elements of their task. But whether any of these heuristics[10] could be labeled as "underlying principles of intelligence" (a primary goal of the science) is a moot point. If the Watson project has revealed any underlying principle of intelligence, it just might be: "use massive parallelism to try everything that might work as long as you can attach a confidence rating, and then sort out the best at the end".

Or more cynically, "try anything that might work and choose the outcome that most of these individual alternative computations agree on". In one fell swoop, the difficulties of logic-based confidence assessment described above have collapsed to simply counting alternative outcomes — and maybe this is just what our brains do.

Here is another diagram of the earlier fictitious Watson example. This time it illustrates a simple counting approach to generating the confidence ratings of alternative answers.

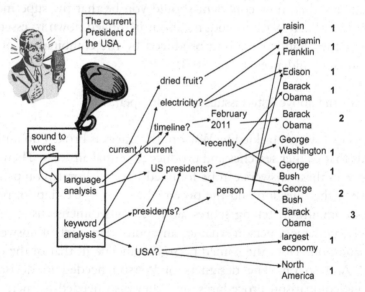

In this small example, "Barack Obama" emerges as the most confident answer simply because it is in the majority. It occurs in three answers whereas the next most confident answer is "George Bush" which occurs just two times. Simple majority voting does the trick[11].

The beauty of such a strategy is that when generating thousands of alternative answers (and not just the measly few I illustrate), most of the really tricky issues of intelligent processing can be avoided. It probably doesn't matter a great deal how you choose to quantify probabilities as long as you do it sensibly. It probably doesn't matter how exactly you construct chains of reasoning, as long as you do it sensibly. Why? Because when dealing with very large numbers of answers the odd, crazy result — like "raisin" above — is expected to be swamped by the sensible answers which will emerge in large numbers from a variety of alternative paths of reasoning[12].

It's the democratic process objectivised. Every member of the population (which are the individual threads of reasoning) has one write-in vote that is worth the same as every other thread's vote. The candidate answer with the largest number of votes wins. The various rogue votes to be expected from a variety of lunatic fringes have the same weight as any other votes, but they do not garner enough support to have a final effect. In this analogy, the winning candidate is the one that is most widely supported by the electorate, the overall system composed of many parallel processes, the individual voters[13].

If our brain cheats just like this we might have to elevate the terminology from "cheating" to something more edifying. Perhaps our brains, which do seem to go in for parallel processing on a grand scale, support a cognitive architecture in which many different parallel paths tackle each problem, and some consensus is extracted at the finish — perhaps nothing more sophisticated than a majority vote. As usual, we just don't know.

But consider a brain-scan of such an architecture — the results might look very like the nanobot scan of our original brain module with multiple outputs instead of just one. No single thread of reasoning is associated with a confidence assessment. And no single thread of reasoning is prominent as one that leads to the eventual conclusion. The crazy (such as, "currant" leading to the "raisin" answer) and the potentially promising (such as, "US presidents" leading to the "George Bush" answer) get equal resources. This would be hard to understand on a thread by thread analysis. Rationality emerges only in the totality of all the parallel threads — there will be many more threads of reasoning leading to the "George Bush" result than to "raisin" (and even more leading to "Barack Obama", one hopes).

Notice also that this is yet another type of holism. The sole reason why "Barack Obama" is the final answer generated is because, within all the answers generated, it is the most frequent one. A holistic strategy that, once again, is unmysterious, indeed utterly simple. Yet it is likely to be invisible to any analysis that is based on a few individual threads of reasoning plucked from the mass of parallel alternative threads.

Either of my proposed confidence-rating strategies is easily illustrated and explained, and each can be summarised in terms of a general principle. In the "number of arrows" strategy the principle might be:

> *The confidence attached to potential answers is determined by the number of directly supportive partial conclusions derived from the query.*

For the "counting" strategy, the principle might be:

> *The confidence attached to each answer is given by the proportion of instances of each different answer that emerges within the full set of alternatives generated.*

And in both cases, a strategy for resolving ties (i.e., more than one answer emerging with equal confidence) would need to be included. One possibility is for rough and ready confidence assessments, one for each thread of reasoning, to bias the simple counting towards higher-confidence outcomes.

Assessment of confidence based on amount of supporting evidence or simply counting final outcomes is not very tricky, is it? (And readily amenable to dynamic visualisation as is the set of confidence-rating procedures Watson used.) Yet the BBC "science" programme made no mention of the crucial parallelism, let alone probed the principles that might underpin the essential confidence ratings of alternative potential answers. The presenter talked instead of nebulous "horizontal" information searching to be contrasted with the usual "vertical" style of computers. Whereas direct reference to the parallelism of Watson's searching strategy (using something like either of the above diagrams) would have given substance to the "horizontal" claim as a parallel set of the normal "vertical" searches.

Similarly, there was no mention that Watson's language processing was optimised to deal with exactly (and only) *Jeopardy*-type queries (with two tricky categories omitted). Instead, we were misinformed that Watson "understands" the nuances of English by means of some unexplained "learning". The truth is that Watson was trained on a set of typical *Jeopardy* examples to optimise the *Jeopardy*-playing process, like the MLP "learning" described in detail earlier. It may well have required innovative and creative thinking (i.e., real intelligence) to set up the training regime, but only within the limited orbit of one-shot, function optimisation[14]. This would result in precious little "understanding" of language within the Watson system (and I'm probably being over-generous here, however we might choose to interpret "understanding" in this context). Watson is an elaborate stimulus-response system — one of four types of *Jeopardy* query in and appropriate answer out — nothing more.[15]

It is obvious that the media, even a BBC "science" programme, must put emphasis on capturing an audience as well as presenting the science. This grabbing of audience interest is easily done with the "amazing Watson" defeating the best humans in a televised intelligence test — our genetic imperative fires on all cylinders, so to speak. Overall the programme did present intelligence as an enduring puzzle, but why omit mention of major limitations and grossly overstate accomplishments?

More puzzlingly, why omit all mention of Watson's possibly crucial contributions to furthering our understanding of intelligence — massive parallelism and confidence calculations? How is the goal of presenting science served by misrepresenting the Watson achievements, and omitting its essential characteristics? It's baffling, and perhaps best explained by incompetence with respect to the particular sciences within the decision-making creatives (and use of front-man scientist from another scientific discipline did nothing to put the brakes on the runaway fiction). The TV programmers are sold on celeb-led programmes, and if it's science then "let's get a known scientist" must have been the cry, with little or no appreciation for inter-disciplinary ignorance.

The scientific papers written by the experts, the Watson creators, do not do mislead in this blatant manner. But, as discussed earlier, when the supposed theory is embedded in a large computer program, there is considerable scope for claims that are arguable because **no significant computer program is completely understood.**

The ultimate explanation is the computer program that generates the behaviour to be explained. But a large computer program is an effective explanation for nobody because the masses of fine detail are too vast and too intricate to take in as a whole. The behaviour of the program must be summarised but to what degree? Smaller summaries are generally easier to understand but can only be obtained by compressing and neglecting large amounts of the program detail. Clearly this detail is needed (otherwise why is it in the program) and some of the omitted detail may be unrecognised as crucial to certain aspects of the behaviour supposedly being explained. The above maxim implies that this will always happen.[16]

Apart from inadvertent misrepresentation, why would a scientist tend to overstate his accomplishments? Science, too, has an audience, and moreover it has an audience that needs to be similarly engaged. Projects need funds, and funding needs friends. A breakthrough, even the promise of a breakthrough, is going to garner more of both than a stark dead end[17]. Science is surprisingly like show business in that it's as much who you know as what you know that helps to attract vital financial support[18].

Ultimately, the proof of the science will be in the fruits it produces. It is, of course, too soon to expect any Watson fruit (apart from the hopeware variety), but surely the general principles upon which Watson's success is based were devised before the Watson experiment was constructed? It is not clear that this happened. The Watson project was not an experimental test of theoretical principles. It was a one-off computer program designed to play *Jeopardy* (minus two question types) at the level of the best humans.

So the Watson project was not primarily science, it was an engineering research project. But now that it has succeeded, we should be able to extract some underlying principles, and test them in further experiments. I can speculate about one of the clear principles epitomised by Watson — massive parallelism.

Watson's central contribution to furthering our scientific understanding of intelligence may be the effective management of parallel processing of alternative threads of reasoning. This, in turn, may point to an effective explanation in terms of the brain-like holistic models.

It is true that any parallelisation buys nothing more than time — a sequential equivalent is always possible but slower. It is also true that an intelligent response must sometimes be timely. The *Jeopardy* game emphasises the crucial nature of speed, but intelligence more generally is intimately bound up with timely responses. This means that parallelisation (especially on a massive scale) could be a salient feature of intelligence.

Quite apart from supporting fast-enough responses, parallelisation opens the door to less complex decision-making. Instead of choosing between or even assessing the relative merits of alternative choices, all options can be pursued (without losing sight of our undeniable truth when exponential increase in alternatives appears).

In pursuit of this parallelisation possibility, it might be productive to further develop the two (fictional) many-threaded parallel processing strategies illustrated above. In a loose sense of the word, these are both "averaging" procedures — a single overall result is extracted from a mass of results that individually can tell us almost nothing about the particular phenomenon being studied. Hence, they are also examples of holistic processing.

Averaging evidence has long been managed by physicists to provide a scientific understanding of, for example, the gas laws — e.g., the amount that nitrogen gas will expand as its temperature is increased at a certain pressure. For a gas composed of millions of individual molecules whizzing about, the prospect of tracking and monitoring individual gas molecules is not a productive option. A comparable absence of viability attaches to the possibility of monitoring individual neurons.

However, the success of the scientific understanding based on so-called "statistical mechanics" relies on two characteristics of gases that our brain's neural networks do not exhibit: (1) each gas molecule can be treated as equivalent to every other molecule; every molecule makes exactly the same contribution to the way a gas behaves, e.g., its pressure increases by a predictable amount when heated; and (2) gases exhibit very few scientifically interesting behaviours in comparison to brains. This profligacy of brains undermines the chances of pinning down specific, cognitively-meaningful patterns in the oceans of poorly circumscribed and equivocal brain-scan evidence despite the possibility that characteristic (1) is exhibited by all the neurons of certain regions of the brain.

Notwithstanding these extra difficulties, in the cognitive-level models outlined above we do have a large set of alternative parallel answer threads, and we've got a (probably) more massive set of parallel neuron pathways in the underlying brain model. But remember the attractive isomorphism between our earlier models, apparent close similarities that came to naught despite our best efforts to span the abyss.

Taking a new tack, a repetitive theme in the earlier analyses suggests an entirely different speculation as general a basis for understanding intelligence — from a grammatical underpinning to language versus ungrammatical utterances, a tussle between learning and forgetting, and a tension between consistency and acceptable inconsistency in our beliefs, time and again we've seen the notion of compromise arise in our musings about intelligence.

Is intelligence, as an abstract phenomenon, founded upon the distribution of finite resources to satisfy infinite needs? If so, compromise and trade-offs between the various dimensions of need — e.g., verbal communication and visual perception — will be vital. Could compromise, between what's good enough[19] and what's the best we can do, be a key to intelligence? In support of this conjecture we've seen how tricky the idea of ultra-intelligence becomes once optimum performance is contemplated. The conundrums quickly escalate when the notion of super fast and unerringly accurate reasoning is combined with perfect memory.

Here's a compromise proposal: intelligence is fundamentally an optimised set of interdependent compromises. It is founded on memory management competence that permits collective optimisation (of the various dimensions of compromise) determined by the interplay of brain-structure limitations and environmental characteristics — i.e., the nature of the underlying machine, and its survival in context (of Earth rather than, say, a satellite gas ball of Betelgeuse).

In short, is intelligence an optimal organisation of compromises? A state of affairs largely determined by the interaction of the characteristics of the "machine" on which the intelligence program is running, as well as by the nature of the world that it developed within. The individual good-enough points of compromise, which are likely vary from individual to individual, could then account for much of the difference in the cognitive competence found among humans. The more resources applied to, say, language skills for example, the less will be available for application to, say, competence in mathematics.

This general view has a knock-on effect for AI. Consider, for example, the good-enough range of compromise between memorisation and recall (i.e. size of memory and speed of retrieving knowledge). For a network of slow and individually unreliable neurons it is likely to be very different from that of a conglomerate of fast and reliable electronics.

This is just one dimension along which a compromise procedure can operate. Here are some others illustrated cavalierly, i.e., the shapes of the curves are: unknown in detail; likely to vary from person to person; and likely to vary from context to context, as well as with respect to time, for any given person. However, the general trends illustrated are defensible. In all examples, we see increasing effort along the horizontal axis (left to right), and increasing quality, such as more effective reasoning, along the vertical axis (bottom to top).

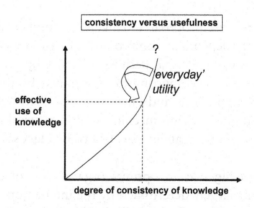

In the above illustration of knowledge management our compromise conjecture suggest that an intelligent system puts just enough effort into checking, tracking down and eliminating inconsistencies as is required to maintain an effectively useful store of knowledge. Then its perceptions, beliefs, plans, statements, etc., will be erroneous on acceptably few occasions which for an evolved intelligence means that in general it survives and prospers, and hence this particular compromise framework is reinforced.

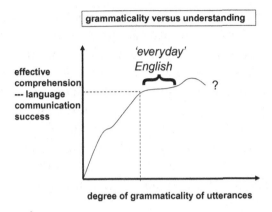

With respect to effective communication, we anticipate that in general the resources committed to spoken grammatical utterances will be as much as is required to achieve good-enough communication. Among the many complexities attaching to language usage there will be, for example, social pressure to vary the effort devoted to grammatical sentences dependent upon the communication context, e.g., a formal lecture versus chit-chat with a close friend.

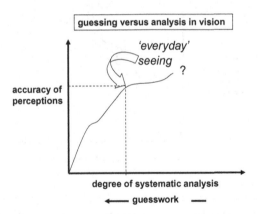

Given such a set of relationships, each admitting a range of "good enough" compromise positions, any integrated collection may be even more variable, but not necessarily. One can imagine that particular choice points on one dimension may constrain the good-enough choice points on another if a similar total resource is efficiently distributed. So a "just good enough" compromise in one dimension may permit a "better than average good enough" compromise in others. However, it is to be expected that overall this set of component aspects of compromise will exhibit variation. But an intelligence that is to communicate with and co-exist with others must, one would think, embody something close to a common set of compromise positions. This would be constrained to be the case for an intelligence that develops within a society of others, but would not necessarily hold true for an alien incomer — a stranger from a strange land, a deep sea dweller, a Martian or an Artificial Intelligence.

Human intelligence is not, despite IQ ratings and their ilk, a one-dimensional phenomenon. Despite an anticipated overall commonality among the compromise structures of a co-evolved group of humans, we all have our cognitive strengths and weaknesses, so here's the resulting picture:

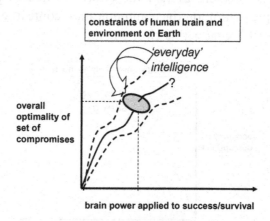

The position of the shaded ellipse represents the general level of intelligence achievable with a human brain on planet Earth. In reality this shaded disc is an irregular multi-dimensional "space" within which the variations of human intelligence exist. It illustrates the scope of compromise structures (i.e., variety of compromise positions along each dimension of compromise) to be found within, say, a specific human culture.

External to the "machine" is the world that the intelligence has evolved within. Perception of external phenomena may always be based on compromise between what is expected, and what external stimuli can be detected and analysed. Particular compromise points will be dictated by relative importance — say, benefits of speed versus cost of being wrong. From an evolutionary standpoint (and I know of no other), the optimal points of compromise will be dictated by what was necessary to survive in the world within which the system evolved, tempered by what is possible in terms of the finite resources and performance limitations of the life-forms' internal chemistry and physics.

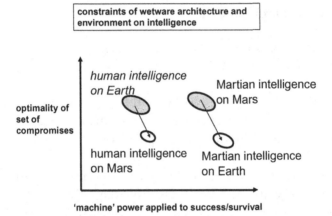

constraints of wetware architecture and environment on intelligence

optimality of set of compromises

human intelligence on Earth

Martian intelligence on Mars

human intelligence on Mars

Martian intelligence on Earth

'machine' power applied to success/survival

In this wholly dreamed-up diagram there is an implication that the overall optimality of the collection of compromises is a rough measure of intelligence. Without venturing to prejudge the cognitive skills or brain power of Martians, it can be taken to suggest that an intelligence that evolves in one environment will not perform as well in another despite extra effort being applied. Any intelligence will necessarily perform less intelligently within an alien environment of comparable complexity.

From this analysis, intelligence stripped of its human bias may be viewed as an interlocking set of compromise procedures whose individual points of good-enough compromise are determined by the operating characteristics of the "machine" available and the nature of the environment within which it has evolved.

With regard to the external world, the compromise conjecture implies that a world within which intelligence can evolve must be consistent enough. Presumably, the laws of physics guarantee a basic consistency, and would in any world. But time pressure, if not inability to know all the relevant information, would often rule out the possibility of "guaranteed" reasoning. To survive, any system will require some degree of guessing in its behaviour. It is the scope and quality of the guesswork that can attain levels where we call it "intelligence". In summary, I am proposing that:

> *The unpredictability of the world makes intelligence necessary and the predictability makes it possible*

This exploitation of a well-behaved world, however, must apply to all life forms, yet the vast majority have never made the great leap forward. There must be more to the explanation of intelligence.

If it's not that, uniquely in the vast gamut of life forms, we humans lucked onto a plate-stacking mechanism (and I've argued that it's not), then what is the answer? Is it memory management and learning where the latter may be rolled into the former? So, is it fundamentally an issue of memory — that humans can store and organise (i.e., learn), interrelate (i.e., reason about), modify (i.e., adjust facts and beliefs) and retrieve vast amounts of information? Is it this knowledge management, which must include compromise optimisation, that accounts for humanity's great leap forward? Maybe, but even if all these hunches are true, it doesn't take us very far forward on our quest for scientific understanding.

It provides an explanation of sorts for our many sub-optimal behaviours, and for variation (both observed within humanity and postulated beyond). But is it a theory? No, not a scientific one because there is no way to test it. Why? Firstly, this compromise proposal has not been refined down to the level of detail that would be needed in order to build a computer-programmed model. Presented as a theory, it falls foul of the unanswerable questions concerning whether a particular computer model does what it's observed to do because of (or in spite of) the "theory".

It might be termed a pre-scientific theory in that it presents the possibility of further refinement to a testable state. We might, for example, develop a mechanism for compromise management and its range of variation in, say, grammatical versus ungrammatical utterances. A sufficiently specific mechanism ought to predict testable (average) behavioural characteristics in response to a controlled variation of circumstances, such as talking casually to close friends versus more formal communication with peers. This would not be easy, but it might be possible.

Even more damning is the "lack of joints" problem — the proposal outlines yet another holistic model in which it is the overall optimal set of compromises that may exhibit indications of intelligent behaviour (e.g., it may pass the Turing Test). A model that can exhibit flexible compromise on any particular dimension will tell us little or nothing about the merits of the proposal unless its underlying principles can be successfully reapplied in other dimensions of compromise.

A final proposal, which we can also discern in the Watson strategy, runs counter to both statistical averaging and compromise competences. It is the view that intelligence is based on an integrated set of specialists — in effect, Minsky's Society of Mind proposal[20]. Intelligence is a general phenomenon manifest as a collection of special cases. In Watson, for example, the designers "use more than 100 different techniques for analysing natural language, identifying sources, finding and generating hypotheses, finding and scoring evidence, and merging and ranking hypotheses"[21].

The history of AI is littered with special-purpose procedures that solve an immediate difficulty but lack general utility. Sometimes called "heuristics", or more honestly "kludges", it is easy to appreciate the attraction of special purpose solutions — the immediate problem is solved through the application of (human) ingenuity and imagination. It is equally easy to appreciate their downside. Their use runs up against all that makes science beautiful — elegance, simplicity and generality of theories. This last property also impinges on potential utility. Much of the power of science resides in the exposure of general principles that can be transferred from their discovery context to explain new phenomena. However, we might note that intelligence might be nothing more edifying than a bag of special-purpose tricks that work well enough for one particular life form on Earth.

We might be tempted to infer that homogeneity of brain infrastructure (one vast neuron network) points to the likelihood of general underlying principles, but of course it does no such thing. Homogeneity of infrastructure of a computer (a network of simple electronic components) has absolutely no hegemony over the simple generality, or the tangle of special mechanisms, that could constitute the huge range of programs it could execute. Rather like the (cognitively) useless information processing principles that we extracted from the small brain module, within the computer there will be similar, relatively universal processing principles at work underlying every program but these will be divorced from the "meanings" of each such program — the familiar, structure-meaning abyss.

So why is intelligence a peculiar phenomenon? We've seen a whole raft of reasons, from our closeness to the subject of study and its apparent seamlessness, to a seductive modelling technology that promises more than it can deliver. The maxims that have been featured here and there throughout this book constitute an attempt to focus on ten basic elements within these three problematic perspectives. We can now integrate all of their individual contributions:

1. A major technical obstacle: **the mind exhibits no joints for science to carve at**.
2. A testing problem: **intelligence admits no simple tests**.

3. A vicious circularity: **understanding based on a single example**.
4. An undeniable truth: **structure does not provide meaning**.
5. A genetic imperative: **if it's like me, it's intelligent**.
6. A logical conundrum: **we cannot prove, but we can disprove**.
7. A sad fact: **failure drives science forward**.
8. A measure of scientific progress: **by their fruits you will know them**.
9. A technological difficulty: **no significant computer program is completely understood**.
10. An unbelievable truth: **no computer will ever count 1, 2,... exponentially many**.

We have computers that chat, or play games, or solve problems. These demonstrations are persuasive. Embody the computer in a robot and its Artificial Intelligence leaps out at us. A genetic imperative has kicked in: **if it's like me, it's intelligent.** This urge appears to be particularly irrepressible when speech is combined with a human look. Science takes a back seat.

Further complicating the obstructive operation of the genetic imperative, we have the trickiness of the topic itself. There is a major technical obstacle to the progress of cognitive sciences: **the mind exhibits no joints for science to carve at.** The "engine" of intelligence, the human brain, does exhibit well-defined components and hence potential joints for science to carve at. There is, however, no known detailed relationship (beyond the usual style of observation that brain component X "is important for", or "key to", some aspect of the intelligent mind) between brain components and perceived aspects of mind, e.g., memory. In addition, study of the concrete object, the human brain, in order to understand the abstract object, the intelligent mind, runs foul of the undeniable truth that **structure does not provide meaning**.

Further complicating the use of classical empirical science methodology (i.e., experiment and test), we have the observation that

intelligence admits no simple tests. This was first made explicit in the Turing Test — only observation over an extended period permits intelligence to be assessed by behaviour in context. Step-by-step accumulation of independent and replicable experimental test results, the backbone of classical science, is not an option when intelligence is the phenomenon being studied.

Realisation of the impacts of the above maxims reveals the surprisingly destructive nature of simple logic. Scientists recognise a logical conundrum: **we cannot prove, but we can disprove.** This leads to the sad fact: **failure drives science forward.** And who wants the disappointment of failing demonstrations to undermine their firm, positive convictions?

Scientists of every persuasion are logically trapped — the most informative thing that they can do with their theory is to show that it is false, and what right-minded person is motivated to work at the destruction of their hard-won brain child? It is a truly perverse situation.

The reason why these two maxims bedevil AI but hardly ruffle the smooth progress of most other sciences is based on a fundamental difference, one that can be crudely summarised as: the AI researcher builds models to demonstrate and test theories, and the classical scientist tests theories on the world as it is. In the former case there is always scope for wondering how demanding the model world was. In the latter case this scope for doubt scarcely exists.

Onto this difference we must add in the repercussions of the maxims so far considered. We have the inability to build and test carved-out components of intelligence. Hence little opportunity test across components of intelligence and by so doing build a case for the generality of whatever principle is being tested. In classical science, say physics or chemistry, a succession of experimental successes across a range of carved-out components adds credibility to the principle being tested. In the sciences probing cognition, a steady accretion of experimental successes appears to be an almost impossible option — neither across components of intelligence (Maxim 1), nor within a limited context (Maxim 2).

Progress in the science of understanding intelligence is even further hampered by the fact that (currently) the human version is the only example we have to study. This drags the investigation into the vicious circularity of **understanding based on a single example** — the scientist has no sound basis for distinguishing the crucial features of intelligence from the incidental ones of the human example. One example hides feature salience, multiple examples can permit winnowing of the crucial from the incidental.

We must also acknowledge the inadequacies of the basic technology for building models that instantiate theories of intelligence — computer programming. As the Internet, personal computers and countless other examples testify, IT systems are truly amazing and have transformed the world we inhabit. But all is not sweetness and light. All the potential interactions between the masses of finely detailed, discrete elements of an IT system are way beyond the scope of human comprehension. When we add in the fact that the smallest overlooked interaction can cause the most massive system malfunctions, our technological difficulty emerges. **No significant computer program is completely understood**[22]. Consequently, no one (not even the system creators) will ever completely understand an IT system that purports to model aspects of intelligence. The system's full scope and limitations will necessarily be a matter for debate gilded by unconscious wishful thinking, as will the precise reasons why the observed behaviours occur. Yet such a model (a computer program) is currently our best bet because it does provide a well-defined and unequivocal basis for discussion of, and further experimentation with, the behaviours observed. A computer model leaves no scope for wishful thinking about what the model will, and will not, do.

The amazing rise in computer power has fuelled disbelief in the truth that **no computer will ever count 1, 2,... exponentially many**.

Escalation in the number of options to be examined, such as all possible patterns of information processing buried in brain-scan evidence, will defeat all conceivable computers. This is because exponential rises in numbers quickly outstrip the number of microseconds since the big bang, or the number of atoms in the universe, etc. This unbelievable truth thwarts a number of superficially plausible strategies for simulating (and in that sense explaining) intelligence — strategies such as exploring all possible chess moves and choosing the best one in order to play perfect chess.

AI's history of one-off successes and dearth of general principles eviscerates the purported science. A general principle should spawn further science, new projects, further experiments — the explicit products of a general understanding. By such means classical science circumvents the obstructions of logic. For all the reasons mentioned above, it is not clear that AI can do this. Unfulfilled expectation of the coming of AI is a regular event, and just like the evidence for the legitimacy of a religious movement: **by their fruits you will know them**.

General principles can be found (and indeed some have been developed above), but they are so general, so divorced from the detail necessary for computer-model building, that they do not provide an immediate contribution to scientific progress. All sorts of fruit may thus be forthcoming. How are we to distinguish "good fruit" from "bad fruit"? The over-general theory fails to constrain the consequent possibilities. What is lacking are specific predictions to experiment with and so test the theories expounded. This devastating weakness is a direct result of our failure to find joints to carve at compounded by the impenetrability of the modelling technology. Who can state with certainty why the computer model exhibited a particular behaviour, and who is to say that that the behaviour exhibited is indicative of intelligence?

This litany of obstacles in the way of a scientific understanding of intelligence makes no mention of many other, possibly bigger, obstructions to scientific progress. I allude to phenomena such as self-awareness, free will and empathy. This investigation has studiously avoided these aspects of intelligence, not because they are unimportant, nor because they are easily dealt with. In all probability their consideration would severely further complicate an already complex situation.

One possibility is that they will emerge once the basic technical aspects have been dealt with successfully. Another is that some measure of basic intelligence (which is all that this study aspires to tackle) can be understood and made manifest in a computer system without the need to directly address these more nebulous aspects of human intelligence.

None of the above-listed difficulties, either individually or collectively, demands the negative conclusion that AI, and hence a scientific understanding of intelligence, is impossible. The conclusion that they do support is that a scientific understanding of intelligence is not yet close, and that intelligence, human intelligence and super-intelligence are three different, obscurely intertwined concepts.

Even with the handy simplification of the CTM conceded and the "more nebulous" aspects of human intelligence discounted[23], science still needs a few good footholds on its subject of enquiry before it can start exercising its awesome power to explain. Thus Crick's and Watson's inspirational revelation of the double-helix structure of DNA provided a firm platform for the foundations of genetic science which then accelerated away. At the moment, cognitive science still lacks any such solid steps from which it can reach out and come to grips with intelligence. Perhaps we must await our Crick and Watson, and until then cognitive science must plod along probing and testing ideas as best it can.

Is the puzzle of intelligence soluble? No one knows. What makes you clever is not magic. The workings behind your cleverness ought to be discoverable by 21st century science, but are they?

We come full circle back to Alan Turing's prescient proposal — the Turing Test. This examination of the human facility with language has (despite its many critics) endured as a yardstick of intelligence in general. It is undeniable that over half a century of scientific exploration of our language capacity has failed to produce models that come close to confusing a human interrogator. This suggests that our language capacity is not a separable chunk of intelligence; it is deeply integrated with all the fundamentals of human intelligence.

Is intelligence an all-or-nothing phenomenon? If so, reliance on luck might be our best hope. Otherwise science is really stuck, and will remain so until scientific methodology develops the tools to interpret and explain the distributed activities that constitute intelligence (if that's what they are).

The choice, crudely put, is between finding the right chunking strategy, or developing an appropriate scientific methodology for holistic-system understanding, or a bit of both.

In sharp contrast to this inability to find guarantees in intelligent behaviour, the phenomena and constituent objects of traditional science have a comforting tendency to always behave in exactly the same way in exactly repeated circumstances — the bases for building guarantees are everywhere. Is this also a crucial element of what science needs and intelligence does not offer?

In the face of all these difficulties perhaps the sensible scientist should give up. We could accept that intelligence is a phenomenon beyond the wit of man. Is there perhaps a logical impossibility in expecting a system (the human mind) to be able to understand itself? But throwing in the towel is not what science does. Big problems are (in the language of "self help") exciting challenges to be tackled. Such a positive stance, however, cannot hide the fact that, so far, science has made little headway in the development of a detailed understanding of human intelligence, let alone intelligence as an abstract concept.

The particle physicist may well have found evidence[24] for the Higgs Boson to substantiate the favoured framework to account for the empirical fact that matter has mass. In sharp contrast, the cognitive scientist is still looking for the basics of a usable framework — the protons, neutrons and electrons of cognition. The possibility of crucial confirmatory evidence in cognitive science experiments or AI models makes no sense at all within the farrago of vague postulates that currently constitute science's understanding of intelligence.

The grand Singularity programme (should it ever materialise) is presented as all benefit and light. The prospect that it would turn out to be utter darkness and disaster from a human perspective is not touched upon. This absence of consideration of the negative possibilities is a worry. However, humanity has much to worry about; "worries" like global warming are probably more likely and a good deal closer. All evidence points to the conclusion that the Singularity is nowhere near, and those so inclined may take comfort from this.

Perhaps we should give thanks for the extreme improbability of the Singularity becoming anything more concrete than a project "well underway" for the foreseeable future. Over the horizon, who knows? But that's what makes science so interesting, and perhaps, in this case, not a little dangerous.

Endnotes

[1] R. Davis wrote this in 1991 in the *AI Magazine*, vol. 12, no. 3, p. 120, and he added, "Daily life is an intractable problem."

[2] One response from Ray Kurzweil in an interview he gave to *USA Today*, dated 2nd February 2011 about an upcoming match between IBM's latest AI supercomputer, Watson, and the two best human players of a US TV question-answering game called *Jeopardy*. Watson did in fact win (see later in chapter).

[3] *Guardian* online 17th February 2011: "IBM computer Watson wins *Jeopardy* clash. Supercomputer outwits US quiz show champions in epic head-to-hard drive battle." Strictly speaking the *Jeopardy* game is not answering questions, but answering with a question to a description, e.g., "The current president of the USA" should elicit the

answer, "Who is Barack Obama", but we'll call it question-and-answer, or query-and-response. This *Jeopardy* oddity is not germane to the issues we discuss.

[4] *The Hunt for AI* was broadcast in the UK on 3rd April 2012 at 21:00 on BBC 2.

[5] In *The AI Magazine* (see following note) we are told, "The *Jeopardy* quiz show ordinarily admits two kinds of questions that IBM and Jeopardy Productions, Inc., agreed to exclude from the computer contest: Audiovisual (A/V) questions and Special Instructions questions. A/V questions require listening to or watching some sort of audio, image, or video segment to determine a correct answer. [Example omitted.] Special Instruction questions are those that are not 'self-explanatory' but rather require a verbal explanation describing how the question should be interpreted and solved. For example, Category: Decode the Postal Codes verbal instruction from host: 'We're going to give you a word comprising two postal abbreviations; you have to identify the states.' Clue: Vain. Answer: Virginia and Indiana. Both present very interesting challenges from an AI perspective but were put out of scope for this contest and evaluation" (pages 62–63).

A further potentially significant boost for Watson was that the "button pressing time" — i.e., the time delay between coming to a decision and depressing a key — which adds (probably significantly) to the human response time, was not also imposed on the computer system.

[6] "Building Watson: an overview of the DeepQA project" by David Ferrucci, Eric Brown, Jennifer Chu-Carroll, James Fan, David Gondek, Aditya A. Kalyanpur, Adam Lally, J. William Murdock, Eric Nyberg, John Prager, Nico Schlaefer, and Chris Welty in the *AI Magazine*, Fall 2010, vol. 31, no. 3. pages 59–79.

[7] This highly contentious event, which involved accusations of cheating and refusal to disclose details, appears to have contributed precious little to the great quest for AI. A substantial reflective article by Garry Kasparov was published in *The New York Review of Books*, pages 2–3, on 23rd January 2010.

I cannot find a single iota of claim that IBM's massive effort and public success with Deep Blue has contributed anything substantive to our understanding of the principles that underlie intelligence. Nothing in the Watson project suggests its impact will be very different.

[8] In reality Watson's make-up is not this simple; it is described as:

"Watson is a workload optimised system designed for complex analytics, made possible by integrating massively parallel POWER7 processors and the IBM DeepQA software to answer *Jeopardy!* questions in under three seconds. Watson is made up of a cluster of ninety IBM Power 750 servers (plus additional I/O, network and cluster controller nodes in 10 racks) with a total of 2880 POWER7 processor cores and 16 Terabytes of RAM. Each Power 750 server uses a 3.5 GHz POWER7 eight core processor, with four threads per core. The POWER7 processor's massively parallel processing capability is an ideal match for Watsons IBM DeepQA software which is embarrassingly parallel (that is a workload that executes multiple threads in parallel)." I don't know what this means but I'm sure it's an awful lot of computer power.

[9] The history of AI is littered with discarded theories for managing uncertainty within a logical framework (fuzzy logic is a well-explored example, an unusual survivor). None, though, have proved to be adequate in the sense of mirroring the human management of reasoning with uncertainty (which is made all the more difficult because we humans appear to operate quite irrationally at times when confronted with uncertain information). In *Not Exactly* by Kees van Deemter (Oxford University Press, 2010) several chapters explore this problem in depth.

[10] Heuristic is the old-fashioned word for special purpose procedures that human creativity and ingenuity have devised to deal with the various tricky issues that always pop up in every computer model that aspires to exhibit intelligent behaviour.

[11] The cynic must wonder about the degree to which this confidence in a given answer could be simply computed as the relative number of parallel answer threads that agree on each different answer. Certainly the live performance of Watson on the TV show is suggestive of this very simple method of confidence rating. An electronic display is provided of the three most "likely" answers, and in the few fractions of the few seconds before a final choice is made, we see the confidence ratings of each alternative move along progressively until a "winner", the most confident answer, emerges. Clearly, this incremental growth in confidence could simply be new alternative parallel threads checking in with one of the three favoured answers — i.e., just counting. That's what it looks like, but then the display could be mocked up for entertainment purposes with little to do with system realities.

[12] This hypothesis may just be my personal contribution to hopeware, but it's a hypothesis that can be explored, which is to be preferred to hand-waving combined with elusive terms such as "horizontal" searches. But note that my description of the proposed procedure contains words such as "sensibly". It is within such innocent words (other common ones are "interesting" and "relevant"), used to pave the road to AI, that the virtual IEDs, which can derail the project, reside. But note that a simple majority-vote procedure does not require any elimination of "non-sensible" threads, nor any computation of confidences, sensible or otherwise (provided Maxim 10 does not intrude).

[13] An obvious objection to this analogy is that the political democratic process has many well-known weaknesses. It may just be the least bad selection mechanism that humanity has devised. But within Watson (or any other AI system), the rigid objectivity of the "voters" may go a long way towards circumventing the weaknesses of the human-based democratic process. But recall Jaron Lanier's objections to "hive mind" decision making mentioned in a previous chapter and fully expounded in *You Are Not a Gadget* (Allen Lane, 2010), although we are only aiming at basic cleverness, not creative breakthroughs.

[14] Intelligence may turn out to be based on function optimisation, but the function is highly multi-dimensional (i.e., there are many aspects of most situations to optimise, not just "speed" and "accuracy"); it is also a "moving" optimum to be chased. None of the current machine-learning algorithms comes close to the flexibility, broad applicability and open-endedness that might begin to match human learning.

[15] It might be argued that humans are also "no more than elaborate stimulus-response systems". Although this rejoinder has some validity, there is, of course, a world of difference between the scope of the stimuli that Watson accepts and the responses it can generate, and the scope of yours.

[16] In 1984 we directly addressed this problem of explaining our computer model based on Hebbian learning. In "Computer programs as theories in biology" published in the *Journal of Theoretical Biology*, vol. 108, pages 539–564, D. Partridge, V. S. Johnston and P. D. Lopez, we presented step-by-step successive omissions and compressions of the program detail and hence a sequence of explanations. This permits the components of an explanation at one level to be referred up to a more generalised viewpoint, or referred down to the finer details from which the chosen-level component was abstracted. In summary, and more accessible, as "What's in an AI program?", pages 112–118, in *The Foundations of AI: a sourcebook* (Cambridge University Press, 1990) edited by D. Partridge and Y. Wilks.

[17] According to the sad fact that **failure drives science forward**, a "dead end" would tell the scientists that the theory investigated is incorrect. Well, it would, if the science were as straightforward as I've presented it, but it's not. Because of the unavoidable complexity of the computer models, one scientist's clear "dead end" is another's evidence for the changes needed to make substantial progress — and there is no objective test to distinguish between these extremes. Such logic-undermining complexity is not, of course, unique to the sciences that grapple with intelligence. The confusions and altercations associated with "cold fusion" science are documented in the Glossary. More recently, the experimental claims that certain sub-atomic particles were observed to travel faster than light (thereby defying one of the most fundamental of scientific beliefs) has recently been discounted as a "wiring error". A workgroup at CERN announced at the Neutrino 2012 conference in Kyoto, Japan, that after repeated tests by both collaborators and competitors, neutrinos don't break the cosmic speed limit.

[18] As in all other walks of life, nepotism and "old boy clubs" run through the science funding committees despite Herculean efforts to provide support solely on the basis of objective merit (for "the impossibility of simple objectivity" in this context, see Endnote 22 to previous chapter).

[19] "Good enough", rather than optimal or even perfect, has long been recognised as a feature of human intelligence. As long ago as the 1960s Herbert Simon gave it the name "satisficing", and in his small book of essays entitled *The Sciences of the Artificial* (MIT Press, 1969) he proposes satisficing (pages 64–65) as the practical necessity when faced with the impracticality of exponentially many possibilities (in effect, our "unbelievable truth", Maxim 10). For us it is a much more widely applicable strategy, e.g., when urgency for a decision overrides time-consuming extensive analysis of the possibilities.

[20] A hopeware-level theory explained at length in *The Society of Mind* by Marvin Minsky (Picador, 1985) and subsequently explored by in many AI projects. Both the

theory and one of its attempted modelling tests were discussed in the previous chapter. The great merit of this proposal is that the scientist is given licence to dream up a new mechanism to solve each new problem that occurs. It's almost an anti-scientific theory, but that doesn't mean it's wrong.

[21] Stated on p. 68 of the *AI Magazine* article referenced in Endnote 6, above.

[22] Towards the end of June 2012 a major UK banking group lost the ability to update customer accounts. It seems that a "patch" (a relatively small piece of programming) was added to the bank's IT system, tested overnight and used the following day. At which point a basic behaviour of the IT system (the updating of customer accounts) ceased working. Nearly a week of top-priority work by the IT experts was required to restore the IT system's lost behaviour. How could this straightforward and fundamental behaviour suddenly become unavailable, and then take days to restore? Simply because no one, not even the bank's IT experts, fully understands exactly how this IT system does what it is observed to do. The IT technologists were all grappling with unmanageable complexity, and that was in the context of relatively simple and well-defined banking procedures.

[23] Both of these basic assumptions are wide open to challenge: the CTM was dealt with at length in an earlier chapter, and there is a wealth of philosophical argument to the effect that what I call the "more nebulous" aspects of human intelligence cannot be conveniently avoided. They are the essences of humanness, and therefore crucial elements of human intelligence. This standpoint, if valid, does not necessarily gut my enquiry into "basic intelligence", but it might.

[24] Notice that the evidence does not (and never can) *prove* the existence of the Higgs Boson, and compelling evidence for this particle can never *prove* the validity of the relevant theory. Good evidence for this particle can only boost confidence that the theory is on the right track as an explanation of matter. The fact that this particle's existence was predicted from a general theory is an example of a strong point in classical science — one that is grounded in efforts to understand the world as is, rather than efforts to construct explanatory models.

Glossary of Maxims

In the search for a scientific understanding of intelligence, the central argument of this book is developed within a framework that derives from a variety of sources. The inescapable quirks of logic, compounded by unhelpful psychological predispositions, are further compounded by technical difficulties seemingly intrinsic to the nature of intelligence. Complicating the science even further, problems inherent in the modelling medium — Information Technology — must be acknowledged and minimised insofar as that is possible; they cannot be ignored or eliminated.

I have distilled this diverse framework down to ten maxims that together form the basis of my explanation for the failure of modern science to understand intelligence. Few, if any, of the ten are original. It is their selection as fundamental, and their interaction, that offers the new perspective that is the body of this book.

Each maxim is explained with examples and, where possible, with reference to other popular-science books that expound further.

First, the technical difficulties that emerge when "intelligence" is the topic that scientists want to understand:

1. A major technical difficulty: **"the mind exhibits no joints for science to carve at".** *

Modern science works on a divide-and-conquer strategy. Complex phenomena are broken down into a collection of more or less independent

---

* Use of Jerry Fodor's brutal metaphor (p. 128, *The Modularity of Mind*), although suggestive of "carve off" rather than the possibly more accurate "carve out", is designed to promote the importance of this maxim.

components. Thus matter is composed of molecules which are composed of atoms which are composed of protons, neutrons and electrons. Each of these smaller problems can be studied in isolation as approximations to their original manifestations within the total complex. The resultant set of understood pieces is then re-integrated by adding back the neglected aspects of their original contexts. In the case of matter this might be inter-atomic and inter-molecular forces. In this way, science has been spectacularly successful at developing an understanding of almost every complex phenomenon it has tackled.

As Fodor succinctly puts it, "The condition for successful science... is... relatively simple subsystems which can be artificially isolated and which behave, in isolation, in something like the way they behave *in situ.*" (*The Modularity of Mind*, p. 128).

The glaring exception is intelligence, and a major technical reason is that no way has been found to break intelligence down into more or less independent components. Science has found no joints that permit a cutting up of intelligence into independent, manageable pieces. Obvious potential "joints" are language, vision, memory, etc. But each of these components of intelligence is a huge and complex problem in itself. Language, for example, cannot be isolated from memory, from beliefs, or from interactions with the world and with other people.

As the argument in this book makes clear, it is in two small books — *The Modularity of Mind* (MIT Press, 1983) and *The Mind Doesn't Work that Way* (MIT Press, 2000) — that philosopher Jerry Fodor stresses the pertinence of this maxim. It has ominous consequences for the relevant sciences: cognitive science, Artificial Intelligence, and the branch of neuroscience whose goal is to elucidate details of the brain's role in supporting intelligent behaviour.

2. A testing problem: **intelligence admits no simple tests.**

Intelligence is notoriously difficult to define and to identify with any certainty. There is no acid test. Moreover, the intelligent response to a given stimulus may be an unintelligent response to the same stimulus in a different context. So, there is no simple option for step-by-step accumulation of independent experimental outcomes to amass support for a theory of intelligent behaviour.

Added to this, a system that learns — which every intelligence must — undermines the possibility of repeating exactly the same experiment. This is to deny another fundamental of scientific practice: namely, the importance of replicating experimental outcomes in order to confirm the outcomes initially recorded.

The Turing Test (which features prominently in this book) is an experiment in which an interrogator engages in text-based "dialogue" with two other "systems" in order to identify which is the human and which the machine. This Test requires a probabilistic judgement based on extended dialogue. Argument rages over the probabilistic nature of acceptable recognition strategies (e.g., that the machine is mistaken for a human more often than pure chance would deliver), and what defines the length of the necessary testing. What is quite clear, though, is that the outcome of only one, or a few simple tests, cannot be sufficient for a definitive judgement, nor can it be an unequivocal contribution towards such a judgement.

The ability of the scientist to replicate experimental results is a fundamental plank of empirical science, but one that is not totally sound. As an example of its vulnerability (which the study of intelligence exacerbates), see the claims and counter-claims surrounding the failures to replicate the "cold fusion" experiments of the 1980s in, for example, *Too Hot to Handle* by Frank Close (Penguin, 1990).

3. A vicious circularity: **understanding based on a single example.**

Until Martians or adventurous denizens of Betelgeuse decide to put in an appearance, Earth science is stuck with but one single example of intelligence to study: the human, brain-based version. So the scientist must wonder, for example, is "forgetting" a crucial feature of intelligence? Is it a universal of intelligence? It is certainly a feature of the human version, but is it the result of an optimal compromise (between memory size, and fast recall and reasoning) that all intelligent systems would need to accommodate? Is it a crucial feature or is it an incidental feature of the human example. Is it a weakness caused by our reliance on slow and unreliable neurons?

A scientific understanding involves sorting out the salient features of the phenomenon under study from the incidental ones exhibited by the particular examples being studied. A scientific understanding of toasting, for example, hinges on recognition of the importance of the reduction of the surface molecules of a slice of bread to produce a hot, carbon-rich surface. This understanding does not involve chromium boxes and electricity, or Boy Scouts with pointed sticks at the ready around a campfire. It is the very existence of several rather different toaster systems that facilitates the discrimination between crucial and purely incidental features of the phenomenon — if the feature is evident in only one example, it cannot be crucial, and conversely, if it is evident in all examples it might well be a crucial, or salient, one.

With just one example to study — the human one — the scientist is trapped in a whirl of unverifiable conjectures: what is key versus what is incidental to the phenomenon of intelligence.

4. An undeniable truth: **structure does not provide meaning.**

The essence of this truth is captured in the down-to-earth observation that *"we can't understand the plot of Downton Abbey by taking apart the television"*. Watch and listen to a television show on your TV screen or on your laptop computer. A good detective story will soon have you conjecturing about "who dunnit" with a variety of possible candidates appearing as the story unfolds.

Now turn off the sound and picture but leave the show running. How will you enjoy the story? Not at all; there will be no meaning for you. The story will have vanished. But inside the machine it's still running, exactly as before, on whichever device you chose to use. So can you look into the device (with whatever probes and instruments you choose) and find out what's happening in this TV show? No, of course you cannot (at least not without merely recreating the sound and pictures, i.e., without exploiting the key bridging features, sound waves and visual imagery). This is because examination of the structure of the show running on some device cannot reveal what's going on in the show, its meaning. In the absence of salient-feature insights there is an abyss separating structure and meaning.

"Structure" here encompasses the dynamics of operations on, or executed by, the bare structures. It is not limited to static forms. The TV programme being watched and the human brain at work are both composed of constantly changing "structures". The fundamental truth captured by this maxim is that an explanation at one level — such as inter-neuron signalling — cannot, on its own, deliver an explanation at a higher level, such as the "thinking" that the dynamic brain structures are supporting.

The impossibility of this leap, summarised as "structure to content", is further emphasised when you realise that exactly the same show will be running totally differently on the two different devices, TV or laptop — i.e., one meaning but based on two entirely different structures. This truth impacts on efforts to understand intelligence by studying the structure and operating principles of the brain. In his classic book, *Minds, Brains and Science* (Harvard University Press, 1984), the philosopher John Searle uses this truth as one of his four axioms that together mean (for him) that Artificial Intelligence is impossible. Here it is used to support the claim that simply probing the workings of a brain is not going to reveal much about intelligence.

More recent, more accessible, and more directly pertinent to the main force of this maxim, the book *The Optimism Bias* (Robertson, 2012) by neuroscientist Tali Sharot, is full of the latest developments in brain-science's understanding of the human mind. As always, they are all couched in indirect and general terminology, such as "key to" or "involved in"; specific structure-to-meaning associations are entirely absent. Encompassing more than the repercussions for brain science and broadly supportive of this maxim is *Brainwashed: The Seductive Appeal of Mindless Neuroscience* by Sally Satel and Scott O. Lilienfeld (Basic Books, 2013).

The following unhelpful psychological predispositions stem from the nature of "intelligence" as a subject of study, and from the oddities of formal logic.

5. A genetic imperative: **if it's like me, it's intelligent.**

When we observe an entity that speaks, reasons, or otherwise behaves rather like a human, we immediately "see" intelligence at work. In addition, if the entity also looks a bit like a human then our judgement with regard to its cognitive level receives a further positive boost. This drive in humans to "see" intelligence in systems that bear some resemblance to ourselves may well be an inescapable propensity. One that has evolved over countless years of encountering fellow humans as the only systems that behave rather like us.

The recent advent of computer systems has destroyed the dependability of this hitherto "sure bet". Just as the modern world has undermined the validity of other genetic predispositions (see the example of sugar, below), it has also resulted in an over eagerness to "see" intelligence at work in sophisticated computer systems. This seriously undermines efforts to understand intelligence, especially if a face and some limbs are added to the computer system — i.e., a robot.

The book, *Why We Feel* (Helix, 1999) by bio-psychologist Victor Johnston, argues for the emergence of a variety of humanity's genetic predispositions (although not this one) as necessary for human survival on planet Earth. The sweet taste of sugar, for example, although the basis of a killer-urge in modern man, developed as a vital genetic pre-disposition to enhance survival when sugars were scarce throughout our long evolutionary history. What was once survival-enhancing now has just the opposite effect. In exactly the same way, although far less of a life-and-death issue, we can view our propensity to "see" intelligence at work on the flimsiest evidence.

6. A logical conundrum: **we cannot prove, but we can disprove.**

Proof and refutation, the certainty that some assertion is true or that it is false, are the two sides of logic. "All swans are white" is either true or false. But there is no symmetry between the two outcomes — a single counter-example (a non-white swan) is enough to falsify this potential truth, but there is no number of positive examples (white swans) that will confirm a general truth.

A single scientific experiment may disprove a theory. Endless experimental outcomes may support a general theory but they will never prove its truth. In summary, proof of a general theory or hypothesis by means of experiments is impossible whilst disproof is easy. This latter possibility, potential disproof, is often adduced as the defining characteristic of a scientific proposition.

This lack of symmetry between proof and refutation is seldom grasped by non-scientists, and often glossed over by the scientists themselves. Historian Rebecca Stott[†] talked of Darwin needing to "prove his theory of natural selection". Well, it's still not proven, and never will be. What Darwin did was amass evidence to support his theory and, more importantly, he found no evidence that refuted it. It is not the failure to find fossil evidence of many transitional species (so-called

---

[†] UK BBC Radio 4 *Start the Week,* 9 am, Monday 9th April 2012.

"missing links") that threatens a refutation of evolutionary theory. The key to its continued viability as science's best explanation of the variety of life-forms on Earth is that there have been no discoveries of fossil species wildly out of place — e.g., no rabbits put in an appearance before all fish.

It is the failure of evidence to support a theory that really matters. The strength of evolutionary theory rests on the many decades of unearthing fossils, none of which seriously challenges the accepted succession of life forms. Just one fossil bunny pre-dating all fish would send shock waves through the faith of all evolutionists. The rational faith of the scientist should be open to objective refutation (in contrast to the irrational faith of the religious believer).

Karl Popper's book *The Logic of Scientific Discovery* (English edition, 1959) goes through this logical conundrum at length. The poverty of a purely logical analysis of the scientific process as conducted by humans (with unavoidable bias and prejudice) is expounded by Thomas Kuhn in his classic book *The Structure of Scientific Revolutions* (University of Chicago Press, 50th anniversary edition 2012).

7. A sad fact: **failure drives science forward.**

A consequence of the imbalance between true and false is that only experimental failures provide the scientist with clear guidance — i.e., the theory that engendered the experiment is wrong. A successful experiment only tells the scientist that the relevant theory may be true. Buried among many other postulated differences between scientific and everyday thinking in *The Unnatural Nature of Science* (Faber and Faber, 1992), the developmental biologist Lewis Wolpert notes "the preference that people — including scientists — have for trying to confirm hypotheses, rather than for trying to refute them". It is this preference that, in the special case of Artificial Intelligence, triggers the genetic imperative to "see" general success in the smallest example. This unholy alliance provides the (apparently) inexhaustible fuel to subvert the scientific process — decade after decade of delusion about being on the edge of the necessary breakthrough.

8. A measure of scientific progress: by their fruits you will know them.

This maxim is the test for a scientific breakthrough, and St Matthew appears to be the first person to get this observation on the record hundreds of years ago. He is credited with declaring: "Beware of false prophets, who come to you in sheep's clothing... You will know them by their fruits... every good tree bears good fruit... Therefore by their fruits you will know them." (*The Bible*, New King James Version, Matthew 7: 15–20).

Scientific understanding of phenomena is founded on extracting the general principles that underlie particular examples. For example, neon and the other so-called inert (or noble) gases are all classified as inert because, following the principles of complete and incomplete electron orbits, their atoms all exhibit complete electron-orbit structures. The same principles explain why other elements are not inert, as well as the specific and various ways that they are not inert, e.g., what other elements they will combine with, and how they will combine. The electron-orbit principles also explain neon light (the reason why neon can be induced to emit light as well as the reason for the colour of that light) as well as many other phenomena.

The generality of a principle is established by further successful applications of the principle in a variety of new demonstrations. Thus scientific study must be closely linked to general principles as embodied in a theory or hypothesis. The success of this science can then be measured by the range and variety of subsequent phenomena that the new theory can explain and the new experiments or demonstrations that it can suggest to further probe its general validity. A truly general theory will, by definition, give rise to a range of new insights — the fruits of real science. By way of contrast, experiments and demonstrations that are not testing general principles, and consequently do not lead to further somewhat different, new demonstrations are not stepping stones to scientific understanding.

In an update of his 1972 book, *What Computers Still Can't Do* (MIT Press, 1992) and also with his brother Stuart in *Mind over Machine* (Macmillan, 1986), Hubert Dreyfus pointed out the sterile reality of supposedly successful AI projects; it was called "the first-step fallacy". These books document in detail many examples of early AI, the grossly over-inflated claims, succinctly labelled "hopeware". Despite a deluge of criticism (some technically correct) from the AI establishment, the general thrust of this analysis has stood the test of time.

Models used to explore theories of intelligence are computer programs, a modelling medium based on Information Technology (IT). This technology, which is fundamental to the exploration of theories of intelligence, contributes two very different maxims that both relate to misconceptions about the power of IT.

9. A technological difficulty: no significant computer program is completely understood.

Modern programming technology is well-defined in principle but ill-defined in practice. The unmanageable complexity of any significant computer program (i.e., a program beyond a few lines of code) derives from the interactions of many fine-grained decisions, coupled with the fragility of a discrete technology, i.e., the smallest change can have the biggest repercussions. So, no one (not even the programmer) can be sure that they fully understand how the program works, or precisely why it does whatever it is observed to do. This (largely unacknowledged) truth continues to wreak havoc with the schedules and budgets of all large IT systems[‡]. This difficulty is the focus of *The*

[‡] In May 2013 the UK's BBC announced the abandonment of a many-year IT project to digitise its video and audio archive. The cost of the failed project was declared to be £100 million.

Seductive Computer: Why IT Systems Always Fail (Springer, 2011; pages 301–2) in which I put the blame squarely on the nature of programming technology. A 12-point summary lists the full range of fundamental difficulties inherent in this technology:

1. A technology that offers high-precision, but hides the demand for an impossible level of attention to detail.
2. A technology that (falsely) promises a route to the stars [(almost) unlimited possibilities and perfection] for those that are creative enough to devise it.
3. A technology that makes no attempt to reveal the essential dynamics of the routes to its emergent behaviour.
4. A technology that invites "try it and see" (as a seemingly costless and intensely private strategy) — a sure route to eventual chaos.
5. A technology that, like so many drugs, initially delivers easy "highs" (of working programs), but quickly reduces to no more than a never-ending struggle for progressively more-elusive minor successes.
6. A technology that includes no requirement for ease of human comprehension, neither in the mechanisms of its products nor in the patterns of their behaviour.
7. A technology that allows, but offers no inducements, to human-engineer programs.
8. A technology that imposes no sanctions on wholly misleading (stylistic) human engineering of programs.
9. A technology that holds out the promise of formal guarantees for system behaviour, but not seriously.
10. A technology that results in IT systems that can only be tested for validity — a process that is often (nearly) information-less and thus effectively endless.
11. A technology that gives rise to products that no one fully understands.
12. A technology that is amenable to small demonstrations that deny all of the above-listed difficulties.

A litany of major IT-system failures, often based on minuscule errors hidden in hundreds of thousands of lines of programming, has been collected together in *The Day The Phones Stopped* by Leonard Lee (Primus, 1992). *The Seductive Computer* details other examples, more UK-based and more modern.

For example, in 2012 the customers of several large UK banks were denied full electronic access to their accounts for several weeks when a small "improvement" was introduced into the bank's IT system. How could an unforced and presumably well-considered change cause such havoc? Easily. The bank's IT experts did not, and could not, fully understand the implications of the change they introduced, and it took them some weeks of doubtless frantic and intense investigation to develop enough understanding to reinstate full customer access.

An even more surprising repercussion of this "difficulty" has emerged within the esoteric realm of mathematical proof. "I believe in a proof if I understand it"[§] is the mathematician's summary of the basis for a valid mathematical proof. The raw, but ill-understood, power of Information Technology has thrown the centuries-old contentment with mathematical proofs into disarray. Who can understand a proof that is a computer program that explores 200,000 possibilities when a minor misunderstanding of any one could invalidate the supposed proof? Exactly this challenge is presented by the computer proof of the famous four-colour conjecture which asserts that for any conceivable map no more than four colours will be necessary to colour the countries such that no two adjacent countries are the same colour. Alan Bundy tackled this new awkwardness in his talk entitled *A Mathematical Dilemma* which was published in 2006**.

[§] See, for example, this statement in *Nature*, vol. 424, pages 12–13 in 2003, and my letter pointing out the problem with this simplistic viewpoint in "Computer's Ability to Verify Proof is an Illusion", *Nature*, vol. 425, 11th September 2003.

** *The Computer Journal*, vol. 49, number 4, pages 480–486.

A non-formal treatment of both issues can be found in my *The Seductive Computer* (Springer, 2011), Chapters 6 and 8.

There's a fundamental clash between this maxim and the AI endeavour as presented in this book, i.e., constructing a computer system that is intelligent amounts to an understanding of intelligence. The resolution is to be found in the unattainability of absolute certainty. Uncertainty about correct headings for every small step in a very long journey is far more serious than uncertainty about the details of a goal that has been attained with a well-defined model. Science will never claim full and complete understanding of intelligence, but an intelligent program will confer precision and objectivity on all aspects of discussion about how intelligence could work. Such a program would provide a sound basis from which many elements of a scientific understanding could be developed.

10. An unbelievable truth: **no computer will ever count 1, 2,... exponentially many**.

As computers have become faster and faster, so the unbelievability of this truth grows — no computer will ever be able to deal with exponentially many items. Processing large numbers of objects (whether websites to be inspected or chess moves to be analysed) is what computers can be amazingly good at. Large numbers are not a problem. However, some numbers get much larger than others. These impossibly huge numbers typically arise whenever all combinations of a collection of items need to be examined.

The famous example is that of the Travelling Salesman Problem, TSP for short: the salesman is given a list of the cities that he must visit. He is also told the distances between all the cities. His task is to find the shortest tour — i.e., the order in which he should visit these cities so that he travels the least distance. Easy. Examine all possible orderings, calculate the distance to be travelled for each such tour, and choose the shortest. With just 6 cities to visit, there are 720 different possible tours that he must choose from. This is quite a large number but any computer could calculate the length of each of these tours and find the shortest in the blink of an eye.

Now suppose that our salesman is required to visit 30 cities. The same simple process can be used to find his shortest tour, can't it? No, it cannot. Why? Because there are more than 100 million trillion trillion different possible tours of 30 cities. This a huge number, 10 followed by 31 zeroes, or 10^{32} when written in exponential form. But computers are lightning fast, and are only going to get much faster, so what's the problem? Suppose we devote a million computers to this 30-city TSP, each capable of examining a million tours every microsecond. How long will it take to find the shortest tour? With 31 million seconds in a year, our 1 million computers will take 100 million trillion microseconds which is 100 trillion seconds, and so they will need to compute for just 30 million years!

The move from 6 cities to 30 results in a move in computation time from a fraction of a microsecond to 30 million years. It is that sort of jump to the impossible that is unbelievable when casually considering these combinatorial tasks that so quickly grow out of control. It is so-called exponential growth that quickly leads to these impossibly large numbers, and the concise exponential representation (such as 10^{32} above) that further fuels the misconception that some computer, at some point in the future (if not today) will be able to handle the very large number of items generated.

David Harel's small book *COMPUTERS LTD What they really can't do* (Oxford University Press, 2000) is a concise, eye-opening account of this unbelievable truth; it is also a technically sound explanation. Quoting a 1984 *TIME* magazine cover story that claimed "there are no limits on what you can do with software", Harel's response is, "Wrong. Totally wrong."

The impact of this unbelievable truth — summarised as "1, 2,... exponentially many" — is not that intelligence cannot be "captured" in a computer program, but that the straightforward brute force approach (examining all possibilities) cannot be the strategy for its capture. Despite their massive parallelism, brains must be operating more cannily (and so must any computer model regardless of whether we want to model the structures of human intelligence, or simply reproduce, or surpass, human intelligence by whatever means we can). It is these "short cuts" that science must discover. And the hope must be that a few general principles underlie the short cuts, not a large collection of specific tricks.

Index

Printed in the United States
By Bookmasters